Finance and Sustainable Development

There are many studies confirming the relationship between financial systems and economic development, but there are few which examine the degree to which financial systems a) impact the quality of information, b) influence sound corporate governance, c) ensure effective mechanisms of risk management, d) mobilize savings and e) facilitate trade. In the context of sustainability, there should also be a line of inquiry into how a particular financial system influences the assurance and implementation of sustainable development principles and goals.

This book delivers a methodological approach to designing and assessing sustainable financial systems. It provides an original contribution by prioritizing ESG factors in the decision-making process of financial institutions and identifying their impact on sustainable financial systems. The author argues that to achieve financial stability, it is necessary to have in place mechanisms designed to prevent financial problems from becoming systemic and/or threatening the stability of the financial and economic system, while maintaining (or not undermining) the economy's ability to sustain growth and perform its other functions.

The book primarily takes a simulation and experimental approach. It is the first book to take such a comprehensive look at sustainable financial systems as opposed to sustainable finance in general. It will appeal to academics, students and researchers in the fields of economics, finance and banking, business, management and political and social sciences.

Magdalena Ziolo is Associate Professor of Banking and Finance at the Faculty of Economics, Finance and Management, University of Szczecin, Poland.

Routledge International Studies in Money and Banking

For more information about this series, please visit: www.routledge.com/
Routledge-International-Studies-in-Money-and-Banking/book-series/
SE0403

Finance and Sustainable Development

Designing Sustainable Financial Systems

Edited by Magdalena Ziolo

LONDON AND NEW YORK

First published 2021
by Routledge
2 Park Square, Milton Park, Abingdon, Oxon OX14 4RN

and by Routledge
52 Vanderbilt Avenue, New York, NY 10017

Routledge is an imprint of the Taylor & Francis Group, an informa business

British Library Cataloguing-in-Publication Data
A catalogue record for this book is available from the British Library

Library of Congress Cataloging-in-Publication Data
Names: Ziolo, Magdalena, 1976– editor.
Title: Finance and sustainable development : designing sustainable financial systems/edited by Magdalena Ziolo.
Description: Abingdon, Oxon ; New York, NY : Routledge, 2021. | Includes bibliographical references and index.
Identifiers: LCCN 2020037741 (print) | LCCN 2020037742 (ebook)
Subjects: LCSH: Finance. | Sustainable development.
Classification: LCC HG173 .F48757 2021 (print) | LCC HG173 (ebook) | DDC 332—dc23

LC record available at https://lccn.loc.gov/2020037741
LC ebook record available at https://lccn.loc.gov/2020037742

ISBN: 978-0-367-81976-7 (hbk)
ISBN: 978-1-003-01113-2 (ebk)

Typeset in Sabon
by Apex CoVantage, LLC

Contents

Tables

Figures

Contributors

Editor

Magdalena Ziolo, PhD hab, is an associate professor at the University of Szczecin, Poland. She graduated from the University of Szczecin (master's degree in management and marketing and a course in financial management in business entities) and completed three years of doctoral studies. Her research and teaching scope of interest are finance and banking, especially sustainable finance and green banking, sustainable development and public finance. She has received scholarships from the Dekaban-Liddle Foundation (University of Glasgow, Scotland, 2013), Impakt Erasmus + (Ulan Bator, Mongolia, 2017) and CEEPUS (University of Prishtina, Kosovo, 2015, 2016, 2017, 2018). In 2015 her paper (as a co-author) won the Best Paper Award in the Conference on Financial Safety organized in Warsaw by the Social Academy of Science. She is an international member of the State Quality Council, Kosovo (2018–2023). She serves as a reviewer for several national and international publications and regularly attends international scientific conferences in her fields of research. Her scientific achievements encompass over 100 reviewed papers, including academic books (as editor, author or co-author).

Contributors

Iwona Bąk, PhD, is an associate professor at the West Pomeranian University of Technology Szczecin, Poland. Bąk is an expert in the field of quantitative methods, specializing in analyses regarding the use of quantitative methods in economic research, with a particular emphasis on the labour market and tourism, as well as regional development. Bąk is the author or co-author of over 100 articles published in scientific journals from the JCR list and also has practical experience in the implementation of projects carried out on behalf of public institutions.

Iustina Alina Boitan is an associate professor at the Department of Money and Banking, Faculty of Finance and Banking, within Bucharest University of Economic Studies. She is a member of several professional bodies, such as the Financial and Monetary Research Center (since 2008) and the Monetary Research Center within the University of National and World Economy, Bulgaria (research fellow since 2015). She has performed several documentation visits for sharing best practices in teaching and research (Complutense University of Madrid, Spain, 2014; University of Florence, Italy, 2018; University of Geneve, Switzerland, 2018; University of Zagreb, Croatia, 2019; Vienna University of Economics and Business, Austria, 2019) and ERASMUS+ stages (University of National and World Economy, Bulgaria, 2015; Tashkent Financial Institute, Uzbekistan, 2017; University of Dubrovnik, Croatia, 2018; Poznan University of Economics and Business, Poland, 2019). Her research interests focus on financial regulation and supervisory policies, banking systems' efficiency and competition, assessment of banking systems' distress, quantitative supervisory tools (such as the development of early warning systems), ethical or socially responsible banks, national promotional banks, crowdfunding, social and financial inclusion, sustainable development, migration and labour market risks. She has been a member of various research projects obtained through national competitions or funded by the business environment.

Daniel Cash is a lecturer at Aston University. He researches the credit rating industry exclusively, inclusive of other rating systems. He is the author of two monographs on the subject, two forthcoming monographs and the forthcoming *Regulation and the Global Financial Crisis: Impact, Regulatory Responses, and Beyond*, an edited collection.

Katarzyna Cheba, PhD, is an associate professor at the West Pomeranian University of Technology Szczecin, Poland. Cheba is an expert specializing in analyses regarding the use of quantitative methods in economic research, with a particular emphasis on international comparisons in the area of sustainable development, competitiveness of the national economy and regional development, with experience in working with advanced statistical packages STATISTICA, R program, etc. Cheba is the author or co-author of papers published in scientific journals from the JCR list and a member of many projects carried out on behalf of public institutions.

Manipadma Datta has been in management teaching, training and research for more than three decades in India and abroad. His areas of expertise are varied, ranging from sustainable development, climate finance, responsible investments, integrated reporting, circular economy finance and weather derivatives to public management, CSR, strategic management, sustainability and creativity, valuation of intangibles and so on. Currently he is serving as the vice-chancellor of TERI SAS and is also a professor in the Business and Sustainability Department of the university. Prof. Datta has served many institutions of eminence, including the University of Calcutta, Vidyasagar University, Nirma University, Institute of Management Technology Ghaziabad, Indian Institute of Management Lucknow, International Institute of Management Delhi and Vrije University Brussels. He is a regular contributor to national and international refereed journals and is on their editorial boards. He has co-authored two books and authored an audiovisual one on strategic finance. He is on the PhD adjudicator boards of several universities in India and abroad. A regular consultant to the IFC, World Bank and Government of India–sponsored research projects, he has developed housing finance models for the poor in Bangladesh and Nepal and a business model for utilizing the biodiversity of the Cooperative Republic of Guyana, South America. He holds a master's degree and a PhD from the University of Calcutta; he is also a fellow (FCS) of the Institute of Company Secretaries of India.

Marek Dylewski graduated from the Faculty of Transport and Connectivity of the University of Szczecin, Poland, in 1991, and obtained a PhD in finance and banking from the Faculty of Management and Economics of Services of the same university in 1999. He worked in different financial institutions, including banks, and also in the public sector in local and regional government units as a treasurer. He was involved in scientific projects regarding: corporate finance, LGU's finance and the international project: The role and the impact of the factoring and commercial finance industry in the European Union as chief of the survey in Poland. Since 2008 he has been employed as an associate professor at the Institute of Finance of WSB University in Poznan. He has authored and co-authored more than 150 publications, including 20 monographs. His research interests focus on corporate finance, LGU finance and financial analysis.

Beata Zofia Filipiak graduated from the Faculty of Economics of the University of Szczecin, Poland, in 1990 and obtained a PhD in economics and financial strategy from the Faculty of Transport and Connectivity from the same university in 1995. She worked in different financial institutions and obtained the qualifications of a tax advisor in 1998. She was involved in 25 scientific projects regarding corporate financial strategies, financial strategies of LGUs and sustainable development and finance. Since 2013 she has been employed as a full professor at the Institute of Finance of the University of Szczecin. She has authored and co-authored more than 200 publications, including 30 monographs. Her research interests span financial strategies, sustainable finance, financial analysis and financial aspects of sustainable development.

Sylwia Henhappel is a PhD student at the University of Szczecin, Poland. Her research interest is focused on innovations, entrepreneurship, crowdfunding and crowdsourcing.

Filip Iorgulescu is a lecturer at the Bucharest University of Economic Studies, works as a senior economist for the Romanian Fiscal Council and is a researcher associated with the Center of Financial and Monetary Research (CEFIMO). He holds a PhD in finance from the Bucharest University of Economic Studies, and between 2014 and 2015 he received a postdoctoral scholarship for studying risk transmission in financial markets. His main research interests and teaching areas focus on sustainable public finance, fiscal policy, financial markets and risk management. He was a visiting researcher at Universidad Complutense de Madrid (2014) and an Erasmus+ teacher at Université d'Orléans (2016), Tashkent Financial Institute (2017) and Université de Poitiers (2018). He has published in *Economic Modelling*.

Agnieszka Majewska, Dr hab, is an associate professor of finance in the Institute of Economics and Finance at the University of Szczecin. Her research interest encompasses methods of risk transfer by using financial instruments, risk analysis (especially risk estimation and risk mitigation strategies), using derivatives in risk management, investments on capital markets, econometrics and methods of derivatives valuation.

Sebastian Majewski, Dr hab, is an associate professor of finance in the Institute of Economics and Finance at the University of Szczecin. His research interest is focused on behavioural finance (especially behavioural investments on capital markets), econometrics, risk analysis and sport finance (in particular the relationship between share prices and social factors and the valuation of specific assets).

Wesley Mendes is an associate professor of business finance. He has academic and professional expertise in corporate governance mechanisms and in corporate and personal finance. Currently, he is actively involved in a graduate program on financial management at one of the most prestigious and accredited business schools in South America, São Paulo School of Business Administration (FGV/EAESP) (AACSB, EFMD and AMBA accredited) at São Paulo/Brazil. He has been a visiting scholar in different institutions, like the University of Texas at Austin and Brigham Young University. His main research interests include funding mechanisms in organizational and individual levels, financial innovations, behavioural finance and corporate governance. He has collaborated with many academic, professional and governmental institutions from Brazil, the United States, Portugal, Italy, the UK, Australia, Mexico, Chile, and France. He also has collaborated with boards of listed and non-listed companies.

Piotr Niedzielski is a professor at the University of Szczecin and the Faculty of Management and at the Pomeranian University in Słupsk. His research areas are economics and transport organization and the management of innovative processes with a particular emphasis on the services sector. His management experience includes receivership manager, member of the board and supervisor boards, head of the faculty, vice-dean, dean and vice-rector. He is a member and manager of teams performing a number of national and international projects, including an investment and development project worth PLN 68 million, "Center for the Transfer of Knowledge and Innovation for the Service Sector – Service Inter Lab" implemented at the Faculty of Management and Economics Services, University of Szczecin. He is the author of over 100 scientific articles in the area of practical economic life.

Daniele Schilirò has a degree in economics from the Catholic University of Milan. He is an associate professor of economics at the University of Messina, Department of Economics. He was the winner of the Bank of Italy Scholarship "Bonaldo Stringher" in 1980. He attended graduate-level courses at the University of Cambridge (England) and Yale University (USA). He is a research scholar and member of the Scientific Committee of the CRANEC, Catholic University of Milan. He has been a research scholar and visiting professor in several universities (University of Cambridge, UK, 1985, 2002; Oxford University, UK, 1991; University of Twente, Netherlands, 2005, 2007; University of Dubai, UAE, 2014; University of Valencia, Spain, 2016 and 2017). He published in high-ranked journals: *Economic Modelling, Journal of Environmental Management, Economia Politica, International Business Research, Journal of Applied Economic Science, Rivista Internazionale di Scienze Sociali*. His main research topics concern industrial organization, microeconomics, behavioural economics, the eurozone economy, environmental economics and economics of growth and innovation.

Ria Sinha has expertise in the areas of climate finance, voluntary sustainability standards (VSS) and business sustainability, with nearly eight years of experience in industry and academia. She holds a PhD in business sustainability from the TERI School of Advanced Studies (SAS), New Delhi, and has a master's in economics from the University of Calcutta. She was an HSBC Scholar in TERI SAS. As a DAAD-German Indian Climate Change Dialogue Fellow, she spent quite some time in Germany and presented her research at conferences held in Vienna, Berlin and Boston, Massachusetts. She was also invited as an ICRIER Young Scholar by the National Bureau of Economic Research (NBER) Summer Institute, Massachusetts. At present Dr. Sinha is conducting a research project with the University of Szczecin, Poland. She is associated with the Energy and

Resources Institute (TERI), New Delhi, as a fellow and also with TERI SAS as a visiting faculty in the Department of Business and Sustainability.

Anna Spoz, PhD, is an assistant professor in the Department of Finance and Accountancy at the John Paul II Catholic University of Lublin. She is a manager and lecturer of the postgraduate programs "Accounting and Tax" and "Management and Finance in Public Administration". She is the author of over 50 scientific publications mainly on finance and accounting.

Andreea Stoian is a professor of finance at the Department of Finance in the Faculty of Finance and Banking from the Bucharest University of Economic Studies – BUES (Romania). She holds a PhD in finance at BUES. Since 2012, she is the head of the Center for Financial and Monetary Research (CEFIMO) at BUES. She was a visiting scholar at the Universite Paris 1 Pantheon-Sorbonne (CES), Universite Libre de Bruxelles (DUL-BEA) and Universidade do Porto (FEP) and an Erasmus guest lecturer at the University of Poitiers, University of Bordeaux and University of Orleans. She is the first vice-chair of the International Network for Economic Research (INFER), co-founder of the Romanian Association of Finance and Banking (RoFIBA) and fellow of the Monetary Research Center at the University of National and World Economy (Bulgaria). She is also the executive editor of the *Romanian Journal for Fiscal Policy* (RJFP). Her main research interests are in fiscal policy, mainly focused on fiscal sustainability and vulnerability, public debt and budget deficit, but also on financial markets and sustainable finance. She is published in *Economic Modelling*, *Eastern European Economics*, *Empirica*, *Journal of European Economics*, *Applied Economics Quarterly*, *Managerial Finance* and *Czech Journal of Economics and Finance*.

Malgorzata Tarczynska-Luniewska, Dr hab, is an associate professor of finance in the Institute of Economics and Finance at the University of Szczecin. Her research interest encompasses the capital market (especially fundamental and portfolio analysis and risk diversification), statistics, econometrics and multivariate and taxonomic methods and application of multivariate and taxonomic methods in economies (in micro and macro scale) and in the capital market.

Acknowledgements

The Chapters 1–10, 12, and 14 are a part of the research project funded by the National Science Centre Poland, grant no. OPUS13 no. 2017/25/B/HS4/02172.

Chapter 1

Introduction

Magdalena Ziolo

Financial systems have been considered so far based on the economic context, inter alia, whether they ensure the safety and stability of the functioning of financial institutions, whether they protect depositors and clients against unfair practices, or whether they adequately secure micro and macro prudential risks. In these aspects, assessment and measurement of the operation of financial systems is carried out with links between institutions and their resilience to shock through stress tests. The 2008 crisis has shown that such an approach does not guarantee a comprehensive approach to risk management because the environmental and social risk is neglected. Finances today have a wider role than those defined in the 1950s aimed at meeting the needs of shareholders. Today's finance addresses primarily social needs and supports solving environmental problems. It is therefore possible to move away from the shareholders' perspective only to the stakeholders'. The perception of the context of values also changes if the finances are to build sustainable value for stakeholders. This forces a new approach and thinking about finances and financial systems that should be designed and monitored for sustainability and not just for stability. Therefore, financial institutions, regulations, and the entire architecture of financial systems must be reconstructed in terms of considering the decisions and strategic documents of key social and environmental risks such as climate change, social exclusion, hunger, poverty, and income disparities. Sustainable financial systems, financial institutions that contain solutions that support decisions and development of financial products and investments that achieve social and environmental goals, and the criterion of profit are not the only deciding factors on the operation of financial institutions.

A literature review provides information on how the financial system can facilitate decision-making on the trade-offs between economic, social, and environmental goals of sustainable development. Analysis of definitions regarding the concept of financial stability indicates the development of this concept under the influence of the global financial crisis. The new paradigm indicates the need to consider risk factors, and in particular in the new paradigm, environmental, social, and governance (ESG) factors take a special place. Moreover, the evolution highlights broadening this conception to a value to society as a stakeholder triple line: people, planet, profit. To achieve financial stability, it is necessary to have in place mechanisms designed to prevent financial problems from becoming systemic and/or threatening the stability of the financial and economic system, while maintaining (or not undermining) the economy's ability to sustain growth and perform its other important functions.

There are many studies confirming the relationship between the financial systems and economic development, but there are few studies examining the degree to which financial systems a) impact the quality of information, b) influence sound corporate governance, c) ensure an effective mechanism of risk management, d) mobilize savings, and f) facilitate trade. In the context of sustainability, one should also add an inquiry on how the financial system influences the assurance and implementation of sustainable development principles.

With the development of sustainable finance we can observe the increasing role of sustainable financial systems and discuss how to design them to achieve better results in financing sustainable development. The UN Principles of the Responsible Investment (PRI) Initiative define a sustainable financial system as a resilient system that contributes to the needs of society by supporting sustainable and equitable economies, while protecting the natural environment. We can observe nowadays that the governments in many countries have taken substantial steps to develop and promote green finance as a crucial part of environmental finance. Therefore the World Bank emphasizes that the Asia-Pacific region is one of the most active in innovations towards a sustainable financial system. The growing number of studies indicates expectations regarding the improvement of social and environmental results over time in the valuation of the company on their markets. This evidence, and the fact that we observe a systematic increase in the costs of social and environmental damage as a result of negative externalities, indicates the need for a strong custody to create sustainable value.

The aim of the book is to deliver a methodological approach for designing and assessing sustainable financial systems. The original contribution and approach presented in this book proposal consist of prioritizing ESG factors, taking into account in decision-making process of financial institutions and identifying their impact on sustainable financial systems and proposing a new approach to assessing and comparing financial systems with a clear division into sustainable and unsustainable financial systems.

The book covers the research gap concerning sustainable finance and sustainable financial systems. A summary precedes each chapter in the book, which is followed by a conclusion. **Chapter 2** aims to draw attention to the significant gap in the existing literature regarding sustainable development (SD) issues. From the perspective of finance, the approaches seem unsatisfactory, with unanswered questions. The rank issues, its strategic dimension, the amount of financial resources allocated to focus on SD, and the identification of the financial phenomena that fall into this category are priorities. The chapter addresses various research questions related to the inclusion of SD's multidimensional, holistic, and long-term perspectives in finance or the necessary changes in the academic curriculum aligned with the concept of sustainable finance that can lead to an increase in the efficiency of SD funding. **Chapter 3** aims to clarify the relationship between different areas that form the overall system of sustainable funding, including the main components of ESG risk. The financial sector is particularly predisposed to EGS risk exposure, which is an increasingly important element considered in the credit risk management process. Therefore, sustainable finance decisions are those that incorporate ESG risk into the decision-making process. Fuzzy cognitive maps were used to identify factors of the greatest importance for sustainable financial systems and to study the relationships between them. **Chapter 4** centers around the topic of sustainable, socially responsible, or green banking, which is addressed from two perspectives, namely the micro level and macro level. The micro-level dimension of the analytical research aims at emphasizing the peculiarities of this business model and the specific sustainable financial products and services developed by the banks in order to diagnose the role played by sustainable banking in the implementation of environmental and social issues. Furthermore, it is comprehensively reviewed the international frameworks, guidelines, and principles for responsible banking activity, issued by various organizations, with the fundamental purpose of assisting banks in screening and financing socially and environmentally sustainable economic activities. To reveal the spread of socially responsible or sustainable banks across European countries, a map has been created for each main international sustainability framework/standard/principle to visually illustrate those European countries' banking systems witnessing the broadest commitment for sustainable financial behavior. **Chapter 5** provides original knowledge about ways the financial institutions monitor and respond to market needs in terms of adapting the financial products and services offered to respond to the needs of market demand. This includes the role of such instruments as customer segmentation and the design of sustainable value in business models and designating sustainable financial products. Changes in the area of sustainable financial instruments are a consequence of changes in the areas of the markets they concern. **Chapter 6** presents challenges and problems that the insurance sector must face in the context of the emergence of new challenges created mainly, but not only, by environmental risk. Environmental, social, and governance issues should be introduced to the insurance business. Therefore, UN environment and insurance supervisors launched the Sustainable Insurance Forum (SIF) in 2016. This is intended to create a global network of insurance supervisors and regulators working together in the area of sustainable insurance. **Chapter 7** aims to discuss the role played by the capital market in promoting sustainable development. The capital market proves to be a very flexible tool that can meet evolving economic, social, and environmental needs. Even though it has been shown that society as a whole has not fully internalized the need for socially responsible behavior, the capital market has done so because it is understood that SD presents exciting opportunities. The adjustment of the capital market to the sustainability paradigm has not stopped at conceptual strategies or approaches, but has led to the creation of new environmental asset classes and innovative funding solutions such as green bonds. In this context, the capital market has become a leading promoter of the structural reform of traditional businesses from carbon-intensive to climate-friendly projects. **Chapter 8** presents the alternative finance market size and structure. Then the definition and the essence of crowdfunding phenomenon are presented. The authors conducted research and confirmed the hypothesis that crowdfunding is an innovation. The last part of the chapter verifies the compliance of the crowdfunding assumptions with the concept of sustainable finance from a microeconomic and macroeconomic perspective. The chapter also contains the Boston Consulting Group (BCG)-type taxonomy matrix for crowdfunding projects and classification of crowdfunding models by financial risk and the value of the raised funds. **Chapter 9** seeks to answer the questions related to sustainability rating agencies. Furthermore, it analyses closer the methodologies employed by these 'agencies' to decipher whether they can truly fulfil the requirement of the mainstream body of investors looking to invest their resources with those who value and promote sustainability in their practices. If they cannot, then who may be able to fulfil that role? What impact may the requirements of the mainstream investing body have upon deciding the trajectory of this Sustainable Rating Industry? **Chapter 10** discusses the role of state and public finances in taking actions to support the implementation of Sustainable Development Goals (SDGs). Therefore, strategies and government policies need to systemically change consumption and production patterns, encourage the preservation of natural endowments, and reduce inequality. The issue focuses on enhancing sustainable financing strategies and investments at both regional and country levels. Thus, environmental taxation will be presented as an instrument to influence and shape the attitudes of companies and households regarding sustainable development, particularly the role of taxes in reducing greenhouse gas emissions and removing inefficient fossil fuel subsidies. In addition, the role of public expenditure in financing investments and technologies conducive to environmental protection and social inclusion will be discussed. **Chapter 11** is designed to conceptually understand the various typologies of sustainable investing and its role in accomplishing the goals of SD. It attempts to explain the drivers, trends, and various evaluation techniques used by investors to conduct research in responsible investing, instruments based on this ideology, and finally the barriers which deter its spread. The chapter emphasizes the modified measures of socially responsible investing (SRI) and responsible investing (RI) assessment methods, as well as project-related risks and the methodology of its evaluation. The innovative financial instruments which are based on sustainable investing and capture a significant market size in the form of listed equity, bonds, hedge funds, and private equity are explicitly illustrated in this chapter. **Chapter 12** presents theoretical aspects referring to the financial system, financial stability, and sustainability. The chapter identifies ESG risk that matters for sustainable financial systems, defines and provides a methodological approach for sustainable financial systems, and provides recommendations for designing a sustainable financial system. **Chapter 13** outlines the various existing SDG control and monitoring mechanisms in India and emphasizes the need for external auditing of sustainability reports which are at present conducted in an arbitrary manner. The findings from this chapter are expected to benefit policy makers, regulators, academicians, and companies. **Chapter 14** presents the scope and manner of the presentation of the sustainable issues in non-financial statements. The objectives of this chapter are as follows: first, and overview of the non-financial reporting, then its historical development will be presented. Finally, this chapter provides an overview of the new trends and challenges of non-financial reporting.

The targeted participants of the book are the leading representatives of academia, practitioners, executives, officials, and graduate students in economics, finance, management, statistics, law, political sciences, etc. Much emphasis is put on academic issues within the field of financial stability, sustainable finance, and SD.

Chapter 2

Sustainable finance
A new finance paradigm

Andreea Stoian and Filip Iorgulescu

Introduction

Scientists, scholars, economists, humanists, industrialists and civil servants from around the world have expressed concerns about the problems facing society as a whole, the disruptions caused by poverty, environmental degradation, lack of trust in institutions, uncontrolled growth of urbanization and job insecurity that are characterized by common technical, economic, social and political elements and which interact with each other. Five decades ago, they warned that if demographic trends, industrialization and pollution, food production and the depletion of natural resources remain unchanged, the limits of growth will be reached in the next hundred years (Meadows et *al.*, 1972). Daly and Farley (2011) further explained this outcome by pointing out that the current economic paradigm is based on the idea of infinite growth, and given that we live in a world with finite resources, this will inevitably prove to be an impossible goal. While capital market economies can sustain the quest for infinite growth a bit longer due to their superior efficiency in comparison to centralized economies, this is only going to delay the eventual collapse of this paradigm. As a consequence, Meadows and colleagues (1972) suggested that we can create a long-term sustainable society if we curb growth and production of material goods in order to achieve a state of global equilibrium between human population and economic activity.

The scientific interest and concerns in the field of sustainable development (SD) have made significant progress in recent years. We are witnessing a growing body of research that brings empirical evidence on how economic agents have recognized, accepted and internalized this paradigm. However, despite this progress, many studies have shown that there are still gaps in the conceptual substantiation and question whether this concept has been understood correctly. The financial market can play an important role in the transition from the traditional paradigm of economic growth to the SD paradigm through its core function of efficient allocation of the financial resources to the most productive investments. However, from the perspective of finance, the approaches seem unsatisfactory, with unanswered questions. Therefore, through this chapter we aim, on one hand, to draw attention to the significant gap in the existing literature regarding SD issues and briefly present the developments that have emerged in the discourse on SD. We consider this approach important because the ambiguities regarding the conceptual substantiation can generate further ambiguities regarding the implementation of the SD paradigm and the way in which the economic and financial mechanisms will adapt to it. On the other hand, we discuss at a theoretical level the transition from the traditional approach to finance to a more holistic one described by the new emerging paradigm of sustainable finance (SF) and how it can contribute to the achievement of the Sustainable Development Goals (SDGs) recently established by the 2030 Agenda for Sustainable Development.

Theoretical approaches to sustainable development

Faced with growing evidence concerning the shortcomings of the existing economic paradigm, the United Nations Conference on the Human Environment in 1972 issued a declaration containing 26 principles on the environment and development known as the Stockholm Declaration (UN, 1973). Among the principles on which the participants agreed were the following: safeguarding natural resources and wildlife; preservation of earth's ability to produce renewable resources; non-renewable resources must be shared and not depleted; pollution must not exceed the natural ability of the environment to clean itself. Ten years after the conference, it has been found that a number of global environmental problems have not been adequately addressed and that the challenges have grown. Therefore, in 1983 it was decided to establish the World Commission on Environment and Development (WCED) known as the Brundtland Commission whose purpose was to create a united international community to promote common sustainable goals, to identify sustainability issues and to formulate solutions and implement them. The 1987 Brundtland Commission report introduced the term *sustainable development* for the first time and defined it as "development that meets the needs of the present without compromising the ability of future generations to meet their own needs" (WCED, 1987). Since then, the definition and the concept itself have been the subject of numerous debates and criticism, and the scholars and researchers have not agreed on a more comprehensive approach and have not yet bridged this gap, leaving questions still unanswered. In the following, we will briefly present some of the criticisms that reveal the main weaknesses identified in the theoretical substantiation of this paradigm.

Shortly after the introduction of the SD definition, Lélé (1991) identified a lack of consistency in its interpretation and showed significant weaknesses in its formulation pointing to an incomplete perception of poverty and environmental degradation and confusion about the role of economic growth. The author expressed concerns that SD was becoming a fashionable concept that no one cares to define. He also pointed out that there were voices who advocated that this concept should not be defined too rigorously, which would have allowed those with irreconcilable positions not to compromise themselves. However, he supported the idea of revealing new understandings of the relationship between social and environmental phenomena and the most rigorous characterization of the concept in order to be neither misinterpreted nor distorted. He explained literally the meaning of SD as the development that can be continued indefinitely or for an implicit period, while 'development' refers to a process of directed changes. He also pointed out that because there is no clear distinction between objectives and means, SD has often been interpreted as a process of change that can be continued forever.

A decade after Lélé's paper, Banerjee (2003) once again drew attention to the weaknesses of the SD approaches. He stressed that this term was introduced in order to address the environmental issues generated by economic growth, but expressed concern about the ambiguity regarding what is sustainable (economic growth, the environment or both), showing that precisely these ambiguities are the subject of debate. Banerjee supported the idea that this concept aims to describe largely the process of economic growth achieved without destroying the environment and that instead of being a major discovery, it is subordinated to the economic paradigm. He stated that, in fact, the term 'development' is nothing more but a new name for 'economic growth', which thus acquires a much greater relevance and importance. Banerjee believed that corporations play a significant role in SD, but the question is whether current environmental practices are compatible with this concept. He also drew attention to the fact that greening the industry is not the same as SD and that although progress has been made in controlling pollution and emissions, it does not mean that these ways of development are sustainable for the planet.

Hopwood et al. (2005) thought that all the supporters of the new paradigm agreed on the need to change society, but that they have not yet concluded and are still debating what tools are required and what actors should be involved in this process, as still there is no unitary view. They showed that there is an even greater confusion because people tend to use the same word to designate a variety of divergent goals and views that involve different methods or paths of achieving SD and identified three approaches of the SD problem. The status quo approach supporters recognize the need for change, but they do not consider that society or the environment is facing insurmountable difficulties and believe that adjustments can be made without fundamentally changing society. According to their view, the solution to SD is economic growth. They warmly welcome the reduced role of governments manifested by decreasing progressive taxation, social protection, increasing privatization and minimizing regulations, considering that the business environment is the determinant of sustainability. The reform approach accepts the existence of imperative problems, criticizing public policies and business administration, but it does not foresee the collapse of the ecological or social system nor the necessity of some fundamental changes. Proponents of this approach believe that the source of these problems is the imbalance between knowledge and information and accept that at some point profound changes in policy and lifestyle will be needed to address the challenges posed by these problems. They focus on technology, science, information, market change and government reform. The reform approach supporters recognize the key role that governments play in moving towards SD. The business sector also plays its role by putting pressure on and controlling government decisions on taxation and subsidy, while focusing on research and information. The transformation approach supporters view the societal and environmental problems as generated by the fundamental features that characterize society and the human relation and interaction with the environment. They consider that reform is not enough and argue that in fact the problems are endogenous to the economy and the current organization of society that has as its primary goal neither the well-being nor the sustainability of the environment. Unlike the supporters of the status quo approach who think the changes should be made at the level of the top-down management of the existing decision-making structures, the transformation approach proponents see the change through political action within and outside the existing structures.

SD operationalization and its progress

Based on decades of work, in 2015, all the United Nations member states adopted the 2030 Agenda for Sustainable Development that provides a plan for ensuring peace and prosperity for people and the planet (UN, 2015). The agenda has at its core 17 broad SDGs that call all countries to action in a global partnership, regardless of their level of development and which should be achieved within the next several years. All SDGs are deeply interconnected, but according to Rockström and Sukhdev (2016), they can be delimited into societal, economic and environmental goals as follows:

- *Societal goals*:
  - Eradicating poverty everywhere and in all its forms
  - Eradicating hunger, ensuring food security, improving nutrition and promoting sustainable agriculture

- Ensuring a healthy life and promoting well-being at all ages
- Ensuring an inclusive and equitable quality education and promoting lifelong learning
- Ensuring gender equality and better promotion of females
- Ensuring access to modern forms of affordable, reliable and sustainable energy
- Inclusive, safe, flexible and sustainable development of cities and human settlements
- Promoting inclusive and peaceful societies, ensuring access to justice for all and building a responsible, efficient and inclusive institutional infrastructure

- *Economic goals*:

  - Promoting sustainable and inclusive economic growth, full and productive employment, ensuring decent work for all
  - Building a flexible infrastructure, promoting inclusive and sustainable industrialization, stimulating innovation
  - Reducing inequality within and between countries
  - Ensuring sustainable production and consumption patterns

- *Environmental goals*:

  - Ensuring access to clean water and sanitation networks
  - Taking urgent actions to combat climate change and its consequences
  - Conservation and sustainable use of oceans, seas and marine resources for sustainable development
  - Protection, restoration and sustainable use of terrestrial ecosystems; sustainable forest management; combating desertification and reducing soil degradation

- *Overall goal*:

  - Strengthening the means of implementation and revitalizing the global partnership in terms of sustainable development

The approach of Rockström and Sukhdev (2016) emphasizes that SDGs must not be pursued independently because the current problems of the world are systemic and deeply interconnected. Thus, the concept of SD is a holistic one in which economy should serve the higher purpose of societal development while respecting the environmental boundaries of our planet. In this view, Norström et al. (2014) recommended that three key elements be considered for establishing effective SDGs: the necessity of an integrated and systemic socio-ecological perspective; the necessity to make a compromise between the scale of the objectives and the possibility of reaching them; and finally, since the implementation of the SDGs may prove significantly disruptive in comparison to the current paradigm, it should take into account existing knowledge about social change processes both at individual and global levels.

Although the progress made in achieving the objectives has been observed since the adoption of the agenda up to now, the estimates from 2019 show that for many of the SDGs, the 2030 targets will not be reached (UN, 2019). For example, even if the level of extreme poverty is constantly declining, it will not reach a degree of less than 3% of the total population exposed to this risk by 2030. Efforts to eradicate hunger and malnutrition, although they have made significant progress in recent decades, show that the number of people exposed to this risk is growing again. In the field of health and well-being in order to meet the objectives set, efforts must accelerate. To ensure quality education, actions need to refocus on improving the outcomes of lifelong learning, especially among women, girls and marginalized people. Numerous issues have been identified in ensuring gender equality that could undermine the achievement of this goal by 2030. Despite progress, data show that a very large number of people still do not have access to safe water or sanitation, and to achieve the goal of access to a sustainable water supply and sanitation system, annual progress must be doubled. The 2017 Report of the Secretary-General of the United Nations highlighted the need to make efforts to ensure access to energy and achieve targets for resource efficiency and renewable energy (UN, 2017). This required significant financial resources as well as the governments' commitment and willingness to implement new technologies on a wider scale. The 2019 report shows, however, that although there are still 800 million people without access to electricity, progress in this area has accelerated. Even though labor productivity has risen and unemployment has returned to pre-financial crisis levels, the global economy is growing at a slow rate and improvements are still needed to increase employment opportunities. Issues of inequality within and between nations are far from being addressed, and despite the fact that the income of 40% of the bottom population has increased, inequality in income distribution persists. Fast-growing cities can continue to pose a threat to habitat sustainability. Materials consumption jeopardizes the objective of ensuring a sustainable consumption and production plan. Consequently, urgent actions must be taken to ensure the need for materials without leading to overexploitation of resources and environmental degradation. More funds are needed to support actions to mitigate the effects of climate change that is occurring at a much faster pace than anticipated. Although global trends indicate progress in protecting terrestrial ecosystems and forests and ensuring biodiversity, the report indicates that there is a high probability that the targets set for 2030 will not be met due to continued land degradation and biodiversity losses.

The projections of the official report are also confirmed by the existing research. Moyer and Hedden (2020) assessed the progress towards the targeted values of nine of the indicators related to six human development SDGs and found that between 2015 and 2030, the world will make very slow progress towards achieving these goals if the current policies continue to be implemented. They indicated major difficulties in achieving certain indicators, such as access to safe sanitation, completion of the upper secondary cycle and underweight children. They also pointed out that 28 vulnerable countries will not reach their targets and that they will need international assistance. In view of all this, Clark et *al.* (2018) talked about a consensus regarding the insufficiency of public funding and the need to direct it to unlock private finances, which can be a solution to these challenges.

Sustainable finance to achieve the SDGs

Faced with the significant challenges posed by the implementation of the SDGs and by the underlying transition to a sustainable economy, it is only natural to ask how the financial system can contribute to this paramount endeavor. Ike *et al.* (2019) support the idea that the private sector is the key factor in achieving the SDGs through corporate sustainability actions, such as cleaner production, provision of decent work and economic growth, but emphasize that it is still unclear how companies can operationalize SDGs through corporate sustainability. They believe that attention should shift from what determines companies to implement corporate sustainability (i.e., economic benefits, agent theory, etc.) to how it can be done.

Fatemi and Fooladai (2013) believe that the source of these ambiguities and divergences can also be an academic problem. In their opinion, the traditional finance approach focuses on maximizing shareholder wealth, and according to the efficient markets hypothesis (EMH), today's share price is the only indicator to observe. EMH largely originated from Hayek's (1945) work, who stated that information-processing capacity is essential for the permanent and optimal reallocation of resources in a changing environment and that the coordination of the economic actions of a large mass of economic agents is crucial. He concluded that this is only possible in a decentralized decision-making system, based on price market signals in a structure of economic incentives and private property rights that exists in a free market economy. An information-efficient financial market can ensure an optimal selection of investments, companies and sectors that carry out activities aligned with the SDGs. This is because, in an efficient market, the price of the financial assets will reflect all the available information (Fama, 1969) and, thus, it will provide correct signals in the allocation process. The financial market will stimulate the acquisition of information that will be perfectly reflected by the equilibrium price of the traded assets (Grossman, 1978), while companies' performance and the changes of the expected profits will be immediately included in the price of shares. Thus, the price mechanism will permanently contribute to the improvement of the financial resources allocation.

However, many voices have advocated for a revision of the traditional approach to finance, pointing out that many issues need to be rethought in greater connection with the core of the SD paradigm. For instance, how finance defines 'value' and the fact that generally accepted valuation models do not take into account all the economic, social and environmental costs and benefits associated with a project may sway the decision makers to choose investments that lead to unacceptable outcomes (Fatemi and Fooladai, 2013). They think that if rational economic agents who produce goods and services in order to maximize their profits have a constraint imposed on them, this forces them to take remedial actions to compensate for the social and environmental impact of their activity; then the model of profit maximization will turn into a constraint maximization problem. Their view is that the model of maximizing shareholders' wealth which externalizes social and environmental costs no longer reflects current trends and needs and should be replaced with a new one that takes into account all costs and benefits. Clark *et al.* (2018) also pointed out that the preference for short-term profitability can undermine long-term investment decisions and companies' willingness to engage in long-term investments required by sustainable development projects characterized by very high capital costs and long-term returns.

In view of all this, Lagoarde-Segot and Paranque (2018) showed that the need for 6 billion euros per year for the next 15 years to decarbonize the economy and limit global warming has generated a response from many financial economists, practitioners and public and private finance organizations, who asserted the need to revise the role of financial activity within the economic system. In their opinion, this revision would involve at least two aspects: the assertion of social primacy over economic and financial objectives and redefining our relationship with the environment, which instead of being considered external, should be seen as part of the whole to which we also belong. Lagoarde-Segot (2019) restates the important role played by the higher education institutions, academic research and finance disciplines in aligning financial institutions and market participants with the long-term decision-making process required to finance the sustainable economy and society. He thinks that the problem is the discrepancy between changes brought by the new paradigm of sustainable finance to the 'finance function' and the methods of investigation used in this field, which are based mainly on empirical realism, while impact investments involve qualitative changes in financial practice that are incompatible with the existing framework. He also points out that economists distinguish between 'weak' and 'strong' sustainability. 'Weak' sustainability implies substitution between natural, human and manufactured capital, in the sense that the depletion of natural capital can be offset by an increase in the human or manufactured capital. 'Strong' sustainability starts from the premise that there is no compromise between the three forms of capital, and they must be included as complements in the production function. However, regardless of the form, Lagoarde-Segot suggests that the principle of sustainability will bring changes in the intertemporal preferences of economic agents, in the sense that current generations will have to prioritize the well-being of future generations and this will involve short-term welfare losses. From the perspective of time preferences in finance, this change would entail investors to use a required rate of return that may be lower than the equilibrium rate proxied by the weighted average cost of capital and thus make it possible to increase the marginal effect of long-term flows on the net present value of the planned investments. Thus, investment decisions will align with sustainability criteria.

Taking into account that the main functions of the financial system (Levine, 2005) aim at mobilizing and allocating financial resources (which are inherently limited) to their most productive use, monitoring investments during their lifetime with the aid of corporate governance and allowing for risk management and diversification, the field of finance finds itself in a unique position to help with the implementation of SDGs. Thus, Schoenmaker and Schramade (2019a) consider that each of the previously mentioned functions of the financial system can be relevant for transitioning to the new paradigm of sustainable finance:

- Due to its central role in providing funds for the economy, the financial system can play a key part in directing resources towards economic activities that are conducted in accordance to the SDGs, accelerating the transition process. Scholtens *et al.* (2008) also emphasize the important role of financial markets in funding SD through socially responsible investing.
- In what concerns the monitoring function, shareholders can use their position to control decisions made by management boards. In addition to that, the modern view on corporate governance is that it should balance the interests of all stakeholders, including the society and the environment.
- Furthermore, substantial risks are related to environmental issues and the implementation of the SDGs, with two main categories being identified by Dafermos *et al.* (2018): physical risks, which stem from climate-related events, and transition risks, which denote the possibility to encounter financial shocks generated by the implementation of government policies aimed at accelerating the transition to a low-carbon

economy. In this context, the assessment and management of risks become key components of sustainable finance, which can stimulate economic agents to take a long-term perspective and become aware of the severe consequences that can occur if they do not take timely measures towards SD.

In a critical assessment of the current investment paradigm, Schoenmaker and Schramade (2019b) observed that it is almost exclusively based on the narrow approach of risk and return, which largely ignores social and environmental issues, and performance evaluation is usually conducted in comparison with the market, proxied through a benchmark index. Although efforts have been made to include environmental, social and governance ratings into market metrics and indices, they are still regarded as supplementary information and not as fundamental criteria for financial models and investment decisions. In response to these shortcomings, Schoenmaker and Schramade (2019b) advocated that investors should pursue long-term value creation by taking into account several principles, such as:

- The inclusion of environmental risks into the risk premiums of traditional financial models.
- Deviating from factor models and market benchmarks with the main focus on analyzing the future prospects of individual companies in search for long-term value creation opportunities.
- Switching from short-term financial performance to long-term financial and extra-financial performance indicators, with the latter category relating to green standards and the implementation of SDGs.
- Assessing the transition preparedness (how fit is a company's business model to successfully adapt to a sustainable economy?).
- Concentrating investments in smaller portfolios (since the benefits of diversification diminish significantly in the case of very large portfolios) in order to allow a better analysis of the companies that are included in the portfolio, as well as a deeper engagement with their management in order to help them become more sustainable.
- Supporting the transition process by also investing in companies that are currently lagging behind in terms of sustainability but are committed to improve in the future.

Analyzing the practical possibilities for implementing such an investment paradigm that would enhance the social and societal role of finance, Schoenmaker and Schramade (2019b) concluded that the means are available, but the financial industry still has a long way to go. They also recognized that a financial paradigm change is a difficult process that requires a mind-set change promoted through education, as well as governance and regulations.

While many of the desiderates of sustainable finance presented in this chapter may seem rather ambitious in comparison to the well-established principles of traditional finance, it should also be noted that the financial system has already taken some steps towards supporting SD. Thus, increasing numbers of investors and financial institutions have embraced socially responsible investing initiatives which aim to promote the transition to a low-carbon economy (especially through portfolio decarbonization, which refers to divesting from carbon-intensive sectors and redirecting the funds to sustainable businesses) while also pressuring the companies into taking action towards SD. However, as pointed out by Scholtens *et al.* (2008), socially responsible investing is also challenging for investors because they are required to expand their purely financial analysis by including social, environmental and ethical issues in order to assess which companies are acting along the lines of SD. In an attempt to help socially responsible investing, financial markets created new categories of sustainable finance instruments such as green bonds, which denote bonds that are issued with a clear commitment for funding environmentally beneficial projects or activities (WBCSD, 2015). Such initiatives are welcomed and supported by many international financial institutions, as green bonds are largely considered to be the most important financial instrument in the transition to a sustainable economy. However, despite significant increases in recent years, sustainable financial instruments still represent just a small fraction of the financial markets. They may be affected by a loss of confidence due to the absence of a clear unified standard for 'green' instruments, and empirical evidence suggests that they do not offer lower funding costs for issuers, being quite similar to conventional securities (Stoian and Iorgulescu, 2019).

Therefore, while such initiatives are encouraging and they show the willingness of the financial system to adapt and transition to sustainability principles, it is clear that they are far from enough and, as suggested throughout this chapter, a much deeper transformation of mind-sets and society is required in order to implement a new financial paradigm centered on sustainability. In this context, Schoenmaker and Schramade (2019a) proposed a three-stage process of transitioning from the current business model, based on a linear approach of extracting resources, production, consumption and disposal, which is assessed using short-term risk and return indicators, to a fully fledged adoption of sustainable finance:

- Stage 1 focuses on restricting access to funding for companies and sectors that have an unfavorable social and environmental impact through divesting and limiting access to loans. If such initiatives gain momentum on the financial markets, it is expected that they will induce the envisaged companies to reform. From the perspective of corporate finance, this stage does not include changes in the business model of companies, which continues to be centered on the maximization of profits and shareholder value.

- Stage 2 proceeds further by pricing the social and environmental impact of decisions made by companies. These negative externalities are then included in the valuation of stocks and bonds, leading to the estimation of an integrated value which consists of financial, environmental and social values. An example of incorporating these three dimensions (people, planet and profits) is the triple bottom line accounting framework (Slaper and Hall, 2011).
- Stage 3 switches from the restriction of unsustainable companies and sectors to a positive approach of selecting and funding only the projects that have the highest potential to generate a positive environmental and social impact, thus promoting sustainable development in the long run. The implementation of this stage depends on the willingness of investors to accept a fair level of returns by giving up a portion of their financial returns in order to receive higher societal and environmental benefits.

This sequential approach of the transition to sustainable finance is complemented by Soppe (2004), which focused on applying the principles of sustainability to corporate finance. Starting from the simplistic framework of traditional finance, which is centered on agent rationality and efficiency, he envisaged a two-step evolution of financial theory. The first step is represented by behavioral finance, which adjusted economic models by considering multiple variations in agents' behavior. However, this approach has the potential to complicate financial models due to the high number of possible outcomes resulting from various behaviors of economic agents. The second and final step is represented by sustainable corporate finance, which focuses on maximizing a multi-dimensional preference function considering the interests of all the company's stakeholders, including the society and the environment. This cooperative and integrative approach is expected to reunite social and ethical values with financial theory.

Conclusions

At the very beginning, when humankind had just embraced the new concept of SD, the question that aroused the most interest was whether economic development and environmental preservation do not contradict each other. During the 1990s, concerns were directed at how to ensure the SD. Nowadays, the states of the world have agreed on 17 guidelines that indicate the paths that we must follow to ensure SD, which are interrelated and clearly articulate the desired outcomes. The objectives are not limited to safeguarding the environment, natural resources and the ecosystems. They have become much more complex and have considered other dimensions of human development, such as gender, income and country inequality; access to education and healthcare systems; people's well-being; and the protection of human rights.

However, there are studies that show the effects of an unsustainable system. For instance, Pimentel *et al.* (2007) estimated that 40% of the deaths recorded at global levels are related to environmental factors associated with pollution, malnutrition and the occurrence of diseases. These dire effects have become even more visible in the context of the COVID-19 pandemic, which showed that medical systems are not prepared to cope with large shocks that are becoming increasingly more frequent (some of them, like weather events, being potentially connected to climate change), while inequality in access to healthcare has an important impact on the health and well-being of people (Wyns, 2020). Additionally, pollution appears to act as a general unfavorable factor for health conditions that may be unrelated to climate change, with a recent study conducted by Ogen (2020) finding out that long-term exposure to nitrogen dioxide (which can trigger conditions such as diabetes, hypertension and heart disease) may also be one of the factors contributing to coronavirus fatalities.

We are a living entity in a continuous dynamic whose values are changing quite quickly, and this guarantees us the legitimacy of the periodic review of the paradigms that govern our decisions and actions. Therefore, the question we can ask today is whether achieving these goals will really lead us to SD. We believe that it is the duty of each generation to set their standards to ensure the perpetuation of a dignified life on this planet.

References

Banerjee, S. B. 2003. Who Sustains Whose Development? Sustainable Development and the Reinvation of Nature. *Organization Studies*, 24(1): 143–180.

Clark, R.; Reed, J.; Sunderland, T. 2018. Bridging Funding Gaps and Sustainable Development: Pitfalls, Progress and Potential of Private Finance. *Land Use Policy*, 71: 335–346.

Dafermos, Y.; Nikolaidi, M.; Galanis, G. 2018. Climate Change, Financial Stability and Monetary Policy. *Ecological Economics*, 152: 219–234.

Daly, H. E.; Farley, J. 2011. *Ecological Economics: Principles and Applications.* Island Press: Washington.

Fama, E. 1969. Efficient Capital Markets: A Review of Theory and Empirical Work. *Journal of Finance*, 25(2): 383–417.

Fatemi, A. M.; Fooladai, I. J. 2013. Sustainable Finance: A New Paradigm. *Global Finance Journal*, 24: 101–113.

Grossman, S. 1978. Further Results on the Informational Efficiency at Competitive Stock Market. *Journal of Economic Theory*, 18: 81–101.

Hayek, F. 1945. The Use of Knowledge in Society. *American Economic Review*, 35: 519–530.

Hopwood, B.; Mellor, M.; O'Brien, G. 2005. Sustainable Development: Mapping Different Approaches. *Sustainable Development*, 13: 38–52.

Ike, M.; Donovan, J. D.; Topple, C.; Masli, E. K. 2019. The Process of Selecting and Prioritising Corporate Sustainability Issues: Insights for Achieving the Sustainable Development Goals. *Journal of Cleaner Production*, 236: 117661.

Lagoarde-Segot, T. 2019. Sustainable Finance. A Critical Realist Perspective. *Research in International Business and Finance*, 47: 1–9.

Lagoarde-Segot, T.; Paranque, B. 2018. Finance and Sustainability: From Ideology to Utopia. *International Review of Financial Analysis*, 55: 80–92.

Lélé, S. 1991. Sustainable Development: A Critical Review. *World Bank*, 19(6): 607–621.

Levine, R. 2005. *Finance and Growth: Theory, Mechanisms and Evidence in Handbook of Economic Growth*, Elsevier: Amsterdam.

Meadows, D.; Meadows, D.; Randers, J.; Behrens, W. III. 1972. *The Limits to Growth*. Universe Books: New York.

Moyer, J; Hedden, S. 2020. Are We on the Right Path to Achieve the Sustainable Development Goals? *World Development*, 127: 104749.

Norström, A. V.; Dannenberg, A.; McCarney, G.; Milkoreit, M.; Diekert, F.; Engström, G.; Fishman, R.; Gars, J.; Kyriakopoolou, E.; Manoussi, V.; Meng, K.; Metian, M.; Sanctuary, M.; Schlüter, M.; Schoon, M.; Schultz, L.; Sjöstedt, M. 2014. Three Necessary Conditions for Establishing Effective Sustainable Development Goals in the Anthropocene. *Ecology and Society*, 19(3): 8.

Ogen, Y. 2020. Assessing Nitrogen Dioxide (NO_2) Levels as a Contributing Factor to Coronavirus (COVID-19) Fatality. *Science of the Total Environment*, 726. https://doi.org/10.1016/j.scitotenv.2020.138605.

Pimentel, D.; Cooperstein, S.; Randell, H.; Filiberto, D.; Sorrentino, S.; Kaye, B.; Nicklin, C.; Yagi, J.; Brian, J.; O'Hern, J.; Habas, A.; Weinstein, C. 2007. Ecology of Increasing Diseases: Population Growth and Environmental Degradation. *Human Ecology*, 35: 653–668.

Rockström, J.; Sukhdev, P. 2016. *How Food Connects All the SDGs*. Stockholm Resilience Centre. www.stockholmresilience.org/research/research-news/2016-06-14-how-food-connects-all-the-sdgs.html.

Schoenmaker, D.; Schramade, W. 2019a. *Principles of Sustainable Finance*. Oxford University Press: Oxford.

Schoenmaker, D.; Schramade, W. 2019b. Investing for Long-Term Value Creation. *Journal of Sustainable Finance & Investment*, 4: 356–377.

Scholtens, B.; Cerin, P.; Hassel, L. 2008. Sustainable Development and Socially Responsible Finance and Investing. *Sustainable Development*, 16: 137–140.

Slaper, T. F.; Hall, T. J. 2011. The Triple Bottom Line: What Is It and How Does It Work? *Indiana Business Review*, 86(1) (Spring).

Soppe, A. 2004. Sustainable Corporate Finance. *Journal of Business Ethics*, 53: 213–224.

Stoian, A. M.; Iorgulescu, F. T. 2019. *Sustainable Capital Market* in *Financing Sustainable Development*, eds., Magdalena Ziolo and Bruno Sergi. Palgrave Macmillan, Cham: Switzerland.

United Nations. 1973. *Report of the United Nations Conference on the Human Development*. Stockholm, 5–6 June 1972, A/CONF.48/11/Rev.1. www.un.org/ga/search/view_doc.asp?symbol=A/CONF.48/14/REV.1.

United Nations, Economic and Social Council. 2017. *Special Edition: Progress Towards the Sustainable Development Goals*. Report of the Secretary-General, E/2017/66˙, 11 May 2017. www.un.org/ga/search/view_doc.asp?symbol=E/2017/66&Lang=E.

United Nations, Economic and Social Council. 2019. *Progress Towards the Sustainable Development Goals*. Report of the Secretary-General, E/2019/68, 8 May 2019. https://undocs.org/E/2019/68.

United Nations, General Assembly. 2015. *Resolution Adopted by the General Assembly on 25 September 2015*. Transforming Our World: The 2030 Agenda for Sustainable Development, A/RES/70/1, 21 October 2015. www.un.org/ga/search/view_doc.asp?symbol=A/RES/70/1&Lang=E.

World Business Council for Sustainable Development. 2015. *Green Bonds 002˚C – A Guide to Scale Up Climate Finance*. www.wbcsd.org/Projects/Education/Resources/GREEN-BONDS-002-C-A-guide-to-scale-up-climate-finance.

World Commission on Environment and Development. 1987. *Our Common Future*. Oxford University Press: Oxford.

Wyns, A. 2020. *How Our Responses to Climate Change and the Coronavirus Are Linked*. 2 April 2020. www.weforum.org/agenda/2020/04/climate-change-coronavirus-linked/.

Chapter 3

ESG risk as a new challenge for financial markets

Iwona Bąk and Katarzyna Cheba

Introduction

More and more institutions have decided to publish reports with non-financial data or integrated reports that contain both financial and non-financial data. It turns out that financial data are no longer sufficient to assess a company's performance, especially its capabilities and potential. A comprehensive, fair assessment of the enterprise contains not only economic and financial indicators but also environmental, social and organizational order data (Sikacz & Wołczek, 2017). Responsible investing means the consideration of factors related to environmental protection, social responsibility and corporate governance (ESG, the abbreviation for environmental, social and governance) when making investment decisions to generate higher risk-adjusted returns, in particular in the long term. In the United States and Western Europe, the inclusion of non-financial data by investors in the investment process is already a standard (Sroka, 2016, p. 51). As research by Ocean Tomo, LLC, shows, in 1975 the value of the S&P 500 index was determined to be 83% on the basis of tangible assets, and in 2015 the value of this index at 84% was determined by intangible assets. In Europe, 80% and on average 70% of investors in the world perceive integrated reports, i.e. those containing financial and non-financial information, as important or key ones to making an investment decision (EY Report, 2015). Reporting of non-financial data is also becoming an important element of periodic reporting, as exemplified by the implementation of the provisions of Directive 2014/95 / EU of the European Parliament and of the council (*Directive*, 2014). The financial sector is particularly predisposed to EGS risk exposure, which is an increasingly important element considered in the credit risk management process. Therefore, sustainable finance decisions are those that incorporate ESG risk into the decision-making process.

This study aims to clarify the relationship between different areas that form the overall system of sustainable funding, including the main components of ESG risk. Fuzzy cognitive maps were used to identify the factors of the greatest importance for sustainable financial systems and to study the relationships between them.

The chapter is organized as follows: the next section describes the origin and the role of ESG factors included in the risk assessment, the third section provides an overview of the literature on ESG factor examination, the section after that contains the scientific material and method, and the final section offers conclusions.

The origin and the role of ESG

The origin of the ESG concept comes from the field of socially responsible investment (SRI), which involves investment strategies covering not only economic aspects but also environmental, social and management issues (EUROSIF, 2014). Investing in ESG has existed since at least the early 1980s in various forms. In the initial phase, ESG was defined as socially responsible investing and funds, such as the SRI fund of Amy Domini (Ribando & Bonne, 2010). Then SRI evolved into ESG and covered a much broader program. Institutions applying good ESG principles reduce their impact on the environment by reducing carbon dioxide emissions and water consumption; they are socially responsible for the treatment of employees and their role in the community; and establish the best corporate governance practices for an independent, fairly paid board that protects the shareholders' rights. The ESG concept is also called the three pillars of sustainable development (Staub-Bisnang, 2012) and has been referred to by many terms such as ethics, greenery, impact, mission, responsibility, social responsibility, sustainable development and values that include strategies (such as environmental and social management criteria and corporate governance) to generate long-term competitive financial returns and positive social effects (USSIF, 2014).

The United Nations Environment Programme Finance Initiative (UNEP FI), founded in 1992 after the Earth Summit in Rio de Janeiro, also encourages a better implementation of the principles of sustainable development at all levels of activity in financial institutions, indicating the need for financial institutions to consider environmental and social factors and corporate governance (ESG factors) in the decision-making process (Zhao et al., 2018).

The global crisis of 2007, which shook the global economy, and in particular financial markets, forced financial markets and companies to rethink systemic risk. Discussions about changes in paradigms in management, the verification of the approach to enterprise value management (value-based management [VBM]) and discussions on the broad spectrum of business opportunities offered by sustainable development have begun and are increasingly becoming the driving force for innovation and the quality measure of a company management system, providing huge business benefits. The awareness of the key role that the investment community and capital market entities must fulfill in solving the most important ESG problems is growing all over the world. Including ESG factors in financial risk assessments is gaining popularity in developed economies. The total amount of managed assets based in the United States using strategies that actively incorporate social, environmental and government issues into investment decisions increased from USD 3.74 trillion at the beginning of 2012 to USD 6.57 trillion at the beginning of 2014 (an increase of 76% in two years). In Europe, they account for around 41% (7 trillion) of all professionally managed assets (EUROSIF, 2014).

Reports published by various institutions emphasize that companies applying ESG principles strive for a balance between the goals of the organization and the expectations of entities participating in their activities in increasingly complex conditions. The effective management of relationships with these entities allows them to manage risk more effectively and use opportunities, and thus ensure better conditions for achieving long-term success (FTI, 2016). Incorporating ESG factors in financial risk assessments is gaining ground across developed economies. The ESG investing market is expanding at a progressive rate in the United States, Canada and Europe, where ESG investing strategies have increased from $589 billion to $945 billion in the period 2012–2014. Institutional investors are seeking higher alpha in investments (Zioło et al., 2020). Institutional investors who decide to include ESG issues in portfolio management do so for a variety of reasons: to maximize financial returns, to act in accordance with personal ethics and for further social purposes.

ESG-related factors can have a significant impact on the value of companies and securities. Informing stakeholders about the results of actions taken in the ESG areas indicates the transparency of reporting companies. Therefore, it can be expected that the demand for ESG data will continue to increase because databases with ESG data can help investors in making investment decisions.

ESG risk is of particular importance for financial institutions, especially banks, in connection with their role as financial intermediaries and capital-raising entities. Financial institutions are important catalysts for supporting economic development (Zioło et al., 2019b). This role must include and integrate the promotion of sustainable business practices that will otherwise fail, which consequently will lead banks and other financial institutions to apply practices that have a negative impact on the environment and society and thus lose opportunities to create new products and services using ESG (Hachigian & McGill, 2012).

Whereas since 1992 the UNEP FI indicates the need for financial institutions to take into account ESG factors in the decision-making process, financial supervisory authorities of various countries are conducting research to answer the question how the financial sector can contribute to sustainable development (Finansinspektionen, 2016). The development of sustainable financing also promotes the emergence of rating agencies dealing with ESG as providers of ESG information and tools to measure the contribution of enterprises to sustainable development. Muñoz-Torres et al. (2018) examined eight ESG rating agencies and found that they use different assessment methodologies to check how sustainable a company is, but regardless of the signals that rating agencies and ESG indexes send to markets, they are an important aspect of SRI.

In 2010, the European Council approved an action strategy for the European Union 2020 entitled "Europe 2020. A strategy for smart, sustainable and inclusive growth ", which will help transform the EU into a smart, sustainable and inclusive economy ensuring a high level of employment, productivity and social cohesion. Europe 2020 puts forward three mutually reinforcing priorities (Europe, 2020):

- smart growth: developing an economy based on knowledge and innovation;
- sustainable growth: promoting a more resource efficient, greener and more competitive economy;
- inclusive growth: fostering a high-employment economy delivering social and territorial cohesion.

The document also indicates challenges related to the financial market, claiming that

> Global finance still needs fixing. The availability of easy credit, short-termism and excessive risk-taking in financial markets around the world fueled speculative behavior, giving rise to bubble-driven growth and important imbalances. Europe is engaged in finding global solutions to bring about an efficient and sustainable financial system.

In its latest report, published in 2018, the Global Sustainable Investment Association (GSIA) reported that the inclusion of ESG in decision-making has increased by 25% over the past two years, and 30% of assets invested worldwide have included sustainability in the investment analysis. This, together with the pressure from shareholders and stakeholders, which still occupies a high place in the decision-making process, indicates that the ESG market is maturing and is sustainable. GSIA also estimated that 30% of assets to invest globally (over USD 20 billion) included sustainability in investment analysis (HSBC, 2018). In addition, the Hongkong Shanghai Banking Corporation Limited (HSBC) claims that the inclusion of ESG factors will become the norm and will influence choices about who the institutions will do business with and how they will invest, whereas risk management is one of the key drivers of ESG investment among pension and state funds. That is why HSBC continues to work for sustainable development in all global companies. In addition, the report indicated that risk management was one of the key drivers of ESG investment among pension and state funds. Other relevant factors include company strategy, stakeholder pressure, regulatory requirements and financial planning. In fact, 50% of issuers and over 60% of investors have an ESG strategy. This mainly applies to investors from Asia, the Gulf States and the United States (HSBC, 2018).

External research companies, such as Sustainalytics and MSCI (2018) ESG Research, collect and report data on ESG practices of thousands of companies and provide ESG assessments of companies against their industry counterparts. This research provides asset managers with insights that help them incorporate sustainability criteria into their strategies and funds (Morningstar, 2017). The American consulting company Mercer studied the impact of climate change on strategic asset allocation decisions. Risklab, a specialized subsidiary of Allianz Global Investors dealing with investment and risk consulting, has expanded its focus to all ESG components in the "ESG Risk Factors in a Portfolio Context" study. Risklab has built upon a framework for modeling ESG risk factors. This model incorporates different types of sustainability risks such as global warming, human rights, bribery and corruption and performs sensitivity analysis of ESG factors on equity returns. Each of the sustainability factors is modeled as a stochastic process, and the analysis is made sector specific (*ESG risk*, 2010).

ESG factors in decision-making: a literature review

The financial sector is particularly predisposed to EGS risk exposure, which is an increasingly important element considered in the credit risk management process. Therefore, sustainable finance decisions are those that incorporate ESG risk into the decision-making process. ESG risk management systems should be based on the best information, including historical data, experiences, feedback from stakeholders, forecasts, observations, expert assessments, etc. (Zioło et al., 2020). Therefore, literature on these issues is becoming increasingly important.

Investments that do not consider environmental factors will pose a threat to the environment, which may cause a public protest and social risk as a result of it. Before large investments are undertaken, especially those related to environmental and social risks, government consent is needed, and this is where management risks arise. A failure to consider these ESG risks may result in a loss of investor reputation, leading to increasing financial risk. Table 3.1 presents a catalog of ESG factors, i.e. risks related to non-financial factors, such as environmental protection, social responsibility and broadly understood organizational governance.

Table 3.1 ESG risk categories

Environmental Risks	• Climate Change
	• Degradation
	• Pollution
	• Scarcity of Resources
Social Risks	• Consumer Rights
	• Health Risks
	• Human Rights
	• Labor Rights
	• Safety Risks
	• Strikes and Uprisings
Governance Risks	• Broken Corporate Structure
	• Government Approval of Corruption
	• Weak Legal Structure
	• The Relations of Government and Stakeholder

Source: Şimşek et al. (2019).

Environmental risk includes all environmental conditions that can prevent the delay, creation or completion of the investment project, as well as additional costs. Environmental sustainability is primarily influenced by pollution, climate change and environmental policies inappropriately applied by the government. Risks that are taken into account in the analyses of institutions conducting investment activities include a significant increase in the prices of raw energy materials, failure to meet the adopted schedule of measures to improve the energy efficiency of a company, exceeding the permissible level of greenhouse gas emissions, limiting access to water, negative impact on the natural environment, taking into account the biodiversity in the place where the investment is located and loss of customers due to the insufficient level of ecological innovation of products/services offered by the enterprise (Śledzińska & Kuchenbeker, 2017).

Social risk includes consumer rights, health risks, human rights, labor rights, security risks, strikes and insurrections. Detailed types of risk include announcement of an occupational strike by trade unions, inability to meet the assumed production level due to the leaving of key employees (and the lack of deputies with similar competences), inability to carry out service or repair activities due to the inability to recruit qualified employees, serious breach of labor standards and treatment of employees by the company's suppliers and conflict with the local community.

Management refers to the processes and procedures that are used to properly manage any organization. Assessed for ESG risk, it covers all aspects of government relations, including government approvals, licenses and concessions, particularly with regard to economic assets. Examples of corporate governance risks include recording losses due to ineffective supervisory oversight of the company by the board, occurrence of employee abuse/conflicts of interest resulting in significant financial losses for the company, deterioration of the company's/brand's reputation due to concealing information on ongoing proceedings, penalties imposed or identified violations of the rules of conduct/requirements/standards, loss of customers due to an ineffective complaint process and leakage and disclosure of information about the leakage of personal data of the company's customers.

Specific ESG-related factors or issues – irrespective of them having an impact on or from the asset – may have a direct or indirect positive (business opportunity) or negative (business risk/threat) impact on infrastructure assets. A positive impact may lead to financial gain, and a negative impact may lead to financial loss (Table 3.2). Institutional investors have different approaches to ESG. Sometimes they are reluctant to include ESG factors in investment management due to practical barriers, such as the cost and complexity of implementing an ESG investment strategy, or behavioral barriers, such as the concern that ESG factors are "non-financial".

The enterprise's consideration of ESG factors in its operations may have a positive impact on its operational and financial efficiency, as well as the increase in value (Kochalski, 2016). This is confirmed by numerous studies analyzing the impact of ESG factors on the efficiency and the value of an enterprise (Aktas et al., 2011; Baron et al., 1993; Cheung et al., 2011; Cornett et al., 2013; Eccles et al., 2013; Godfrey et al., 2009). According to Kochalski (2016) out of the 85 studies he presented, 80% indicate the existence of a positive relationship between taking ESG into account in the operations of enterprises and their effectiveness and increase in value. Only 6% of studies show no such relationship, and nearly 13% show mixed indications. Only one study (1%) showed a negative relationship.

Table 3.2 ESG impact on and from an infrastructure asset

Impact from infrastructure asset	Impact on infrastructure asset
Infrastructure assets can have a positive or negative impact on the surrounding environment and/or society	Infrastructure assets may be positively or negatively affected by its surrounding environment and/or society
Examples: environmental degradation, pollution, improved access to basic services, health and safety for workers, corruption, etc.	Such external impact on the asset is primarily of a physical or regulatory nature
Feedback loops, i.e. a reaction from the surrounding back onto the asset may occur, e.g. tax breaks, or societal backlash such as strikes and boycotts	Examples: floods, droughts (natural), resource constraints, pollution, demographics, riots, regulatory changes, etc.
Financial consequences can be direct or indirect, e.g. via reputational risks	Assets resilient towards external impacts can anticipate, accommodate, absorb or recover from such impacts

Source: Web and Rendlen (2019).

Motives, behavior and features shaping the tendency of mutual fund managers to consider ESG issues in making investment decisions can be found in the work by Przychodzen et al. (2016). The study shows that the propensity to include ESG factors is positively correlated with risk aversion. Other reasons for including the ESG are the pressure from stakeholders, mandatory compliance with the regulatory framework, investor reputation and fiduciary responsibility for asset managers. Kumar et al. (2016) for two years evaluated 157 companies listed on the Dow Jones Sustainability Index and 809 that are not listed there. It turned out that companies that take into account ESG factors generate higher profits and show less variation in their stock market performance than other companies in the same industry. Khan et al. (2016) came to similar conclusions, claiming that companies with better ESG ratings have higher future returns on shares. In turn, Grewal et al. (2016) reported that shareholder activity in sustainability issues has become increasingly common over the years, and the number of applications for disclosure of ESG factors doubled in 1999–2013. In their opinion, the submission of applications by shareholders is associated with subsequent improvement in company results. The relationship between eco-efficiency and financial results in 1997–2004 was also studied by Guenster et al. (2011). According to them,

> the market's valuation of environmental performance has been time variant, which may indicate that the market incorporates environmental information with a drift. . . . Results have implications for company managers, who evidently do not have to overcome a tradeoff between eco-efficiency and financial performance, and for investors, who can exploit environmental information for investment decisions.

Mathematical, statistical and econometric methods are particularly useful in financial market analysis. The results of many studies have confirmed that the use of quantitative tools in the study of ESG factors is beneficial for the analysis of economic and financial conditions of entities. To identify factors that can increase the company's tendency to implement environmentally friendly technologies, e.g. a polynomial logit model is used (Frondel et al., 2007). A multidimensional comparative analysis was used e.g. to assess sustainable development, using a set of different indicators (Moussiopoulos et al., 2010) and in the analysis of the relationship between corporate responsibility (CR) and finance in the context of ESG risk research (Bassen et al., 2006). In identifying variables affecting environmental protection (Inglehart, 1995) and in researching the relationship between financial results (creating business value) and variables related to research on corporate social responsibility, multiple regression models were used. Multicriteria decision-making (MCDA) methods were used to make balanced decisions in financial institutions and enterprises that take into account ESG risk, which allow the comparison of decision variants with different, often contradictory, criteria (Vicke, 1992). A detailed description of these methods can be found, for example, in the following works: Figueira et al. (2005); Mendoza and Martins (2006), Triantaphyllou (2000), Buchholz et al. (2009), Huang et al. (2011). Earlier studies of the authors of this chapter (e.g. Zioło et al., 2019a, 2019b) also focused on examining the role of environmental, social and management factors in the decision-making process when designing sustainable financial systems. These studies used advanced methods of multidimensional data analysis: multicriteria taxonomy, correspondence analysis and multicriteria decision support methods: the methods from the Promethee group. They were used, for example, to build rankings of EU countries describing their advancement in counteracting negative conditions affecting the development of ESG risk. An important element of the research was also structuring the research problem, i.e. identifying the factors that are most important for sustainable financial systems and researching the relationships between them. Such studies were also carried out for the purposes of this work. The details of the research experiment carried out are presented in the following subsections.

A cognitive map as a tool to structure the research problem

Fuzzy cognitive maps (FCMs) were used to identify the factors of the greatest importance for sustainable financial systems and then to study the relationships between them. The course of the research using this method usually involves several steps. The first one is related to supporting the identification of the most important criteria for the conducted research, in this case for the relationships between factors important for the sustainable financial systems. In the second one the relationships between these factors will be indicated. In the literature (Ozesmi & Ozesmi, 2004; Tan & Ozesmi, 2006; Zioło et al., 2019b), a cognitive map, commonly known as a map of associations, is presented as a solution dedicated especially for the humanities. In this method the dependence between conditions of the studied phenomenon is primarily sought. It means that the basis for its creation are causal dependencies (associations) of a complex nature. In this method the direct as well as indirect conditions are examined. The most known applications of FCM are modeling (Tan & Ozesmi, 2006; Salmeron, 2009; Papageorgiou et al., 2009); knowledge representation and management (Taber, 1991; Wei et al., 2008); political and social fields (Andreou et al., 2003, 2005); engineering and technology management (Lee & Han, 2000); and agriculture and ecological modeling and management (Isaac et al. 2009), as well as prediction (Furfaro et al., 2010; Song et al., 2010). The FCM was also applied to identify the relations between the criteria that are most important for the design of sustainable financial systems (Zioło et al., 2019b). The main advantages of this method indicated in the literature are abstraction, flexibility, adaptability and fuzziness (Tan & Ozesmi, 2006). After selecting the criteria of the greatest importance for the dependency model, a correlation matrix between the considered criteria should be estimated. The final results of the FCM method are a cognitive matrix presenting the average assessments of compound intensity having higher-than-average significance and a cognitive factor map. In the research for this purpose the software FCMapper_bugfix_27.1.2016 was used. To analyze the results of FCM, the following characteristics can be considered: the density and type of variables presented on the map. The density (clustering coefficient) of a fuzzy cognitive map (D) is an index of connectivity, which shows how connected or sparse the maps are (Ozesmi & Ozesmi, 2004). In the study using FCM the following clustering coefficients can be calculated:

$$D = \frac{C}{N(N-1)} \tag{1}$$

or

$$D = \frac{C}{N^2} \tag{2}$$

The first one presents the share of connections indicated for the map in the maximum number of connections possible between N variables (Ozesmi & Ozesmi, 2004). The second one is adopted when the analyzed criteria have a causal effect on themselves. If the density of a map is high, a large number of causal relationships among the variables can be observed.

Another important result of FCM which should be interpreted during the study is the type of criteria presented on the map. This type shows how the variables act in relation to the other variables, and it facilitates the understanding of map structure. The type of criteria is defined by the following characteristics:

a) outdegree $[od(v_i)]$, which presents the row sum of absolute values of a criteria in the adjacency matrix:

$$od(v_i) = \sum_{k=1}^{N} \bar{a}_{ik} \tag{3}$$

b) indegree $[id(v_i)]$, which shows the cumulative strengths of connections (a_{ij}) between existing criteria; it shows the cumulative strength of variables entering the variable:

$$id(v_i) = \sum_{k=1}^{N} \bar{a}_{ki} \tag{4}$$

According to these characteristics it is possible to indicate three types of criteria (Ozesmi & Ozesmi, 2004):

a) transmitter criteria (forcing functions, givens, tails), which have a positive outdegree and zero indegree;
b) receiver criteria (utility variables, ends, heads), which have a positive indegree and zero outdegree;
c) ordinary criteria (means), which have both a non-zero indegree and outdegree.

It should be noticed that the total number of receiver variables can be considered as an index of the complexity of the map (Ozesmi & Ozesmi, 2004; Papageorgiou et al., 2009). On the other side, a large number of transmitter variables describes the "flatness" of a cognitive map, where causal arguments are not well elaborated (Salmeron, 2009). Different maps can be compared in terms of their complexity according to the ratios of numbers of receiver (R) to transmitter variables (T) – (R/T). In more complex maps, these ratios will be larger due to their large number of utility outcomes defined on the maps.

Study results

The basis for the selection of criteria of the highest importance for sustainable finance systems were the opinions of six experts (financial directors) representing the banking sector, the enterprise sector and the local government. They were asked to indicate the four most important factors in three dimensions which are usually considered in sustainable financial systems: ESG ones. Compared to previous authors' research also dedicated to identifying and examining the relationship between the most important criteria from the point of view of creating sustainable financial systems (Zioło et al., 2019a, 2019b), current research focuses on factors describing the selected areas of particular dimensions. In the case of the environmental dimension, these are factors describing the actions taken by people to protect the climate and the environment. In the social dimension, factors that can be described as external to the created balanced financial system describing the existing situation in terms of social development include the level of poverty, the level of acceptance and, in principle, the possibility of accepting the balance between life and work, or the level of education of the inhabitants. In the management dimension, these factors concerned those the environmental dimension as well as the activities undertaken to build sustainable financial systems and did not consider the current situation, e.g. in terms of gross domestic capita (GDP) or public debt. The factors were determined based on a review of the literature, taking into account the ESG factors included in the methodology of the sustainable rating agencies (Zioło et al., 2019a, 2019b).

A detailed list of criteria selected by experts and subject to further analysis is provided in Table 3.3.

The task of the experts participating in the study was to determine both the strength of the impact of these criteria on the sustainable finances system and the direction of this impact (positive influence versus negative). The assessments were made by experts on a 5-point Likert scale, where 1 was the smallest influence but positive for the studied phenomenon and 5 – the largest impact – was also positive. And vice versa: the criterion that was assigned, for example, the value -1 had the least negative impact on the planned system of balanced finances and -5 had the largest.

On the basis of the experts' answers, a correlation matrix between the selected criteria was developed. The FCM collective matrix elaborated on an individual matrix of each expert, as presented in Figure 3.1. The map presents mutual relations and directions of influence between particular criteria according to the relationships between 12 criteria selected by experts for the study.

The maps presented in Figure 3.1 can be interpreted as follows:

1   In the case of the environmental dimension (E1–E4), the most important seems to be criterion E2 (environmental policy management), in which connections with all others were identified. At the same time, all these relationships are positive, which means that an improvement in this factor will also result in an improvement in other factors. The relations between the other criteria in this dimension are similar.
2   In the case of the second analyzed dimensions (social dimension), negative relations (broken lines) between the criteria were observed between S1 and S4 as well as S2 and S1. This means that an increase in the S1 factor (poverty and social exclusion), which means de facto deterioration,

Table 3.3 The main ESG factors incorporated by financial institutions in the decision-making process selected during the study

Environment	Social	Governance
Air quality and climate change (E1)	Poverty and social exclusion (S1)	Audit and control system (G1)
Environmental policy management (E2)	Human capital development and training (S2)	Business ethics and transparency (G2)
Waste management reduction policy (E3)	Work condition (S3)	Risk and crisis management (G3)
Energy management (E4)	Work–life balance (S4)	Vision and strategy (G4)

Source: Authors' elaboration based on Zioło et al. (2019b).

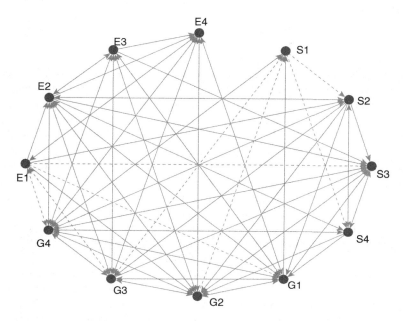

Figure 3.1 The collective FCM of relevant ESG factors
Source: Authors' elaboration.

will negatively affect the S4 criterion (work–life balance). Similar relations apply to the S2 (human capital development and training) and S1 (poverty and social exclusion) criteria.

3   Positive relations between the criteria also occur in the case of the third analyzed dimension (governance). The links are found here between all criteria, which means that an increase (improvement) in each of them will have a positive effect on all others.

4   In general, taking into account all three dimensions, positively linked connections prevail, which means that in most cases, the improvement in individual criteria will also result in improvement in the other ones.

5   Negative relations can be observed between factors belonging to e.g. governance and social dimensions (G2, G3 and E1), which means that an improvement in these criteria (G2 – business ethics and transparency and G3 – risk and crisis management) will cause deterioration within the S1 criterion (poverty and social exclusion), which in this case means an improvement in the social area.

6   In total, the presented map contains 86 connections out of 144 possible and its density is 0.6.

Conclusions

Summing up the results of the study, it can be stated that between the criteria selected by the experts for the study there are mainly positive relationships. Some connections marked as negative actually mean an improvement in e.g. S1 criterion (poverty and social exclusion). All analyzed dimensions are strongly connected with each other. The developed map presents 86 different connections, which experts have identified as significant for a balanced financial system out of 144 possible connections. The developed map presents a rather complicated network of connections between the distinguished criteria. The arrangement of these relations also confirms the observations of other authors (Zioło et al., 2019b), which show that when creating a sustainable finance system, complex relationships between criteria should be taken into account. These criteria should represent different dimensions of sustainable financial systems: ESG ones. It is also worth emphasizing that the presented connections between the criteria are also in line with the assumptions of the latest Strategy for Sustainable Development Agenda 2030, in which, in addition to monitoring progress in achieving individual development goals (17 such goals have been included in the strategy), the relationship between goals is examined. Similar assumptions were made in this study. In addition to the relationships between the criteria within individual dimensions, the relationships between those dimensions were also significant. Most relationships between the criteria were identified for the third dimension – governance (12) versus 7 for the environmental dimension and 8 in the social one. It is also worth emphasizing that the majority of connections between dimensions were also observed in the case of the dimension related to governance (23 combinations of criteria from this dimension with other criteria), and then in the case of environmental dimension (22 connections). By comparison, within the social dimension there were only 11 connections. This means that most interactions are started within the management and environment dimension. Actions taken by decision-makers in these two areas intertwine; in addition, changes in the environmental dimension are strongly dependent on decisions taken within the management dimension.

Based on a literature review, it has been proven that the integration of ESG factors into the decision-making process of financial institutions makes these systems more sustainable. These factors can have a significant impact on the value of companies and securities, and therefore the demand for ESG data will continue to grow because databases with such data can help investors make optimal decisions. Therefore, the inclusion of ESG factors is slowly becoming the norm and has an increasing impact on who the institutions will do business with and how they will invest. In turn, the consideration of those factors in the enterprise's operations may have a positive impact on its operational and financial efficiency, as well as the increase in value (Kochalski, 2016). Therefore, the importance of integrating ESG factors and sustainable development with corporate and investment decisions is becoming increasingly important.

The need for comprehensive research in this area is also becoming increasingly important. The combination of a value-related (quantitative) approach with a qualitative approach makes it possible to search for relationships between determinants that shape financial system stability. In this way, one can answer the question of which factors have a particularly large impact on the stability of this system, taking into account not only direct but also indirect influence. Mathematical, statistical and econometric methods are particularly useful in financial market analysis. The results of many studies have confirmed that the use of quantitative tools in the study of ESG factors is beneficial for the analysis of economic and financial conditions of entities. In this study, the increasingly popular modeling and simulation technique, i.e. FCMs, was used to identify the factors of greatest importance for sustainable financial systems and then to study the relationships between them. They are useful e.g. in administrative sciences, game theory, distributed group decision support systems and for modeling and analyzing economic performance indicators. However, there are not many studies (Zioło et al., 2019b) in which they would be used for research on sustainable finance, and therefore their use in this study seems to be particularly important.

References

Aktas, N., De Bodt, E., & Cousin, J. G. (2011). To Financial Markets Care about SRI? Evidence from Mergers and Acquisitions. *Journal of Banking & Finance, 35*, 1753–1761.

Andreou, A. S., Mateou, N. H., & Zombanakis, G. A. (2003). *Evolutionary Fuzzy Cognitive Maps: A Hybrid System for Crisis Management and Political Decision-Making.* International Conference on Computational Intelligence for Modelling, Control and Automation, Vienna, Austria, 732–743.

Andreou, A. S., Mateou, N. H., Zombanakis, G. A., & Zombanakis, G. (2005). Soft Computing for Crisis Management and Political Decision Making: The Use of Genetically Evolved Fuzzy Cognitive Maps. *Soft Computing Journal, 9*(3), 194–210.

Baron, D. P., Harjoto, M. A., & Jo, H. (1993). The Economics and Politics of Corporate Social Performance. *Business and Politics, 13*, 53–59.

Bassen, A., Meyer, K., & Schlange, J. (2006). The Influence of Corporate Responsibility on the Cost of Capital. http://ssrn.com/abstract=984406/ (15.02.2020).

Buchholz, T., Rametsteiner, E., Volk, T. A., & Luzadis, V. A. (2009). Multi Criteria Analysis for Bioenergy Systems Assessments. *Energy Policy, 37*(2), 484–495.

Cheung, Y. L., Conelly, J. T., & Limpaphayom, P. J. P. (2011). Does Corporate Governance Predict Future Performance? Evidence from Hong Kong. *Financial Management*, 40(1), 159–197.

Cornett, M. M., Erhemjamts, O., & Tehranin, H. (2013). Corporate Social Responsibility and Its Impact on Financial Performance: Investigation of the US Commercial Banks. *SSRN Electronic Journal*, 1–53.

Directive 2014/95/EU of the European Parliament and of the Council of 22 October 2014 Amending Directive 2013/34/EU as Regards the Disclosure of Non-Financial and Diversity Information by Certain Large Entities and Groups.

Eccles, R. G., Ioannou, I., & Serafeim, G. (2013). The Impact of Corporate Sustainability on Organizational Processes and Performance, Electronic Copy. http://ssrn.com/abstract=1964011, www.hbs.edu /faculty/Publication%20Files/SSRN-id1964011_6791edac-7daa-4603-a220–4a0c6c7a3f7a.pdf (15.02.2020).

ESG Risk Factors in a Portfolio Context. (2010). Risklab. www.ipe.com/esg-risk-in-a-portfolio-context/34522.article (15.02.2020).

Europe. (2020). A European Strategy for Smart, Sustainable and Inclusive Growth. https://ec.europa.eu/eu2020/pdf/COMPLET%20EN%20BARROSO%20%20%20007%20-%20Europe%202020%20-%20EN%20version.pdf (15.02.2020).

European Social Investment Forum (EUROSIF). European SRI Study. (2014). www.eurosif.org/wp-content/uploads/2014/09/Eurosif-SRI-Study-20142.Pdf (15.02.2020).

EY Report. (2015). *Tomorrow's Investment Rules 2.0 Report*. EYGM Limited. www.ey.com/.

Figueira, J., Greco, S., & Ehrgott, M. (eds.). (2005). *Multiple Criteria Decision Analysis: State of the Art Surveys*. Springer, London.

Finansinspektionen. (2016). How Can the Financial Sector Contribute to Sustainable Development. www.fi.se/contentassets/123efb8f00f34f4cab1b0b1e17cb0bf4/finansiella_foretags_ hallbarhetsarbete_eng.pdf (15.02.2020).

Franklin Templeton Investments. (2016). www.franklintempleton.pl/investor/ (15.02.2020).

Frondel, M., Horbach, J., & Rennings, K. (2007). End-of-Pipe or Cleaner Production? An Empirical Comparison of Environmental Innovation Decisions Across OECD Countries. *Business Strategy and the Environment*, 16, 571–584.

Furfaro, R., Kargel, J. S., Lunine, J. I., Fink, W., & Bishop, M. P. (2010). Identification of Cryovolcanism on Titan Using Fuzzy Cognitive Maps. *Planetary and Space Science Journal*, 58(5), 761–779.

Godfrey, P., Merrill, C. B., & Hensen, J. (2009). The Relationship Between Corporate Social Responsibility and Shareholder Value: An Empirical Test of the Risk Management Hypothesis. *Strategic Management Journal*, 30(4), 425–445.

Grewal, J., Serafeim, G., & Yoon, A. (2016). Shareholder Activism on Sustainability Issues. 1–56. https://papers.ssrn.com/sol3/papers.cfm?abstract_id=2805512 (15.02.2020).

Guenster, N., Bauer, R., Derwall, J., & Koedijk, K. C. G. (2011). The Economic Value of Corporate Eco-Efficiency. *European Financial Management*, 17(4), 679–704.

Hachigian, H., & McGill, H. S. (2012). Reframing the Governance Challenge for Sus-tainable Investment. *The Journal of Sustainable Finance & Investment,* 2(3–4), 166–178.

HSBC. (2018). Sustainable Financing and ESG Investing Report. www.gbm.hsbc.com/insights/sustainable-financing/sustainable-financing-and-esg-investing-report (15.02.2020).

Huang, I. B., Keisler, J., & Linkov, I. (2011). Multi-Criteria Decision Analysis in Environmental Sciences: Ten Years of Applications and Trends. *Science of the Total Environment,* 409(19), 3578–3594.

Inglehart, R. (1995). Public Support for Environmental Protection: Objective Problems and Subjective Values in 43 Societies. *Political Science and Politics,* 28(1), 57–72.

Isaac, M. E., Dawoe, E., & Sieciechowicz, K. (2009). Assessing Local Knowledge Use in Agroforestry Management with Cognitive Maps. *Journal of Environmental Management,* 43, 1321–1329.

Khan, M., Serafeim, G., & Yoon, A. (2016). Corporate Sustainability: First Evidence on Materiality. *The Accounting Review,* 91(6), 1697–1724.

Kochalski, C. (ed.). (2016). *Green Controlling and Finance.* Theoretical Foundations, CHBeck Publishing House, Warsaw.

Kumar, N. C. A., Smith, C., Badis, L., Wang, N., Ambrosy, P., & Tavares, R. (2016). ESG Factors and Risk-Adjusted Performance: A New Quantitative Model. *Journal of Sustainable Finance & Investment,* 6(4), 292–300.

Lee, S., & Han, I. (2000). Fuzzy Cognitive Map for the Design of EDI Controls. *Information & Management,* 37(1), 37–50.

Mendoza, G. A., & Martins, H. (2006). Multi-Criteria Decision Analysis in Natural Re-Source Management: A Critical Review of Methods and New Modeling Paradigms. *Forest Ecology and Management,* 230(1–3), 1–22.

Morningstar. (2017). The Morningstar Sustainability Rating: Helping Investors Evaluate the Sustainability of Portfolios. www.morningstar.com/lp/sustainability-rating (15.02.2020).

Moussiopoulos, N., Achillas, C. H., Vlachokostas, C. H., Spyridi, D., & Nikolaou, K. (2010). Environmental, Social and Economic Information Management for the Evaluation of Sustainability in Urban Areas: A System of Indicators for Thessaloniki, Greece. *Cities,* 27(5), 377–384.

MSCI. (2018). MSCI ESG Ratings Methodology. www.msci.com/documents/10199/123a2b2b-1395–4aa2-a121-ea14de6d708a (15.02.2020).

Muñoz-Torres, M. J., Fernández-Izquierdo, M. A., Rivera-Lirio, J. M., & Escrig-Olmedo, E. (2018). Can Environmental, Social, and Governance Rating Agencies Favor Business Models That Promote a More Sustainable Development? *Corporate Social Responsibility and Environmental Management,* 26(2), 439–452.

Ozesmi, U., & Ozesmi, S. L. (2004). Ecological Models Based on People's Knowledge: A Multi-Step Fuzzy Cognitive Mapping Approach. *Ecological Modelling,* 176, 43–64.

Papageorgiou, E. I., Markinos, A., & Gemtos, T. (2009). Application of Fuzzy Cognitive Maps for Cotton Yield Management in Precision Farming. *Experts Systems with Applications,* 36(10), 12399–12413.

Przychodzen, J., Gomez-Bezares, F., Przychodzen, W., & Larreina, M. (2016). ESG Issues Among Fund Managers: Factors and Motives. *Sustainability*, *8*(10), 1078, 1–19.

Ribando, J. M., & Bonne, G. (2010). A New Quality Factor: Finding Alpha with ASSET4 ESG Data. *Thomson Reuter, Starmine Research Note*, http://cit eseerx.ist.psu.edu/viewdoc/download?doi=10.1.1.429.577& rep=rep1&type=pdf (15.02.2020).

Salmeron, J. L. (2009). Supporting Decision Makers with Fuzzy Cognitive Maps. *Research-Technology Management*, *52*(3), 53–59.

Sikacz, H., & Wołczek, P. (2017). *ESG Analysis of Companies from the RESPECT Index Based on the ASSET4 ESG Database*. Warsaw University of Life Sciences – SGGW, Warsaw, 158–169.

Şimşek, T., Aydtn, H. I., & Oliwa, B. (2019). Environmental Social and Governance Risk Versus Company Performance, in: Sergi B.S., & Ziolo M. *Financig Sustainable Development*, Springer International Publishing, Switzerland, 249–268.

Śledzińska, K., & Kuchenbeker, J. (2017). *Wpływ czynników środowiskowych, społecznych i zarządczych na sposób funkcjonowania i wyceny spółek*. Deloitte, http://respectindex.pl/pub/Wplyw_czynnik_ESG_sposob_funkcjonowania_wyce na_spolki.pdf (15.02.2020).

Song, H., Miao, C., Roel, W., Shen, Z., & Catthoor, F. (2010). Implementation of Fuzzy Cognitive Maps Based on Fuzzy Neural Network and Application in Prediction of Time Series. *IEEE Transactions on Fuzzy Systems*, *18*(2), 233–250.

Sroka, R. (ed.). (2016). *Non-Financial Reporting. Value for Companies and Investors*. Association of Stock Exchange Issuers, Warsaw.

Staub-Bisnang, M. (2012). *Sustainable Investing for Institutional Investors – Risks, Regulations and Strategies*. John Wiley & Sons, Singapore.

Taber, W. R. (1991). Knowledge Processing with Fuzzy Cognitive Maps. *Experts Systems with Applications*, *2*, 83–87.

Tan, C. O, & Ozesmi, U. (2006). A Generic Shallow Lake Ecosystem Model Based on Collective Expert Knowledge. *Hydrobiologia*, *563*, 125–142.

Triantaphyllou, E. (2000). *Multi-Criteria Decision Making Methods: A Comparative Study*. Springer, Boston, MA.

US Social Investment Forum (USSIF). (2014). Report on US Sustainable, Responsible and Impact Investing Trends 2014. www.ussif.org/Files/Publications/ SIF_Trends_14.F.ES.pdf (23.052016).

Vicke, P. (1992). *Multicriteria Decision-Aid*. John Wiley & Sons, Chichester.

Web, B., & Rendlen, B. (2019). *Guidance Note. Integrating ESG Factors into Financial Models for Infrastructure Investments*. WWF Switzerland, Zürich, 7.

Wei, Z., Lu, L., & Yanchun, Z. (2008). Using Fuzzy Cognitive Time Maps for Modeling and Evaluating Trust Dynamics in the Virtual Enterprises. *Experts Systems with Applications*, *35*(4), 1583–1592.

Zhao, C., Guo, Y., Yuan, J., Wu, M., Li, D., Zhou, Y., & Kang, J. (2018). ESG and Corporate Financial Performance: Empirical Evidence from China's Listed Power Generation Companies. *Sustainability*, *10*(8), 1–18.

Zioło, M., Bąk, I., Sinha, R., & Datta, M. (2020). *ESG Risk Perception in Sustainable Financial Decisions*. Quantitative Methods Perspective, Springer Nature Switzerland AG, Switzerland.

Zioło, M., Filipiak, B. Z., Bąk, I., Cheba, K., Tirca, D. M., & Novo-Corti, I. (2019a). Finance, Sustainability and Negative Externalities. An Overview of the European Context. *Sustainability*, *11*, 4249, 1–35.

Zioło, M., Filipiak, B. Z., Bąk, I., Cheba, K., Tirca, D. M., & Novo-Corti, I. (2019b). How to Design More Sustainable Financial Systems: The Roles of Environmental, Social, and Governance Factors in the Decision-Making Process. *Sustainability*, *11*(20), 56041, 1–34.

Chapter 4

Sustainable banking systems

Iustina Alina Boitan

Sustainable banking: concept, business model peculiarities and sustainable financial products

Mainstream banking is witnessing a broad and long-lasting reshaping of its traditional way of doing business through the incorporation of environmental and social assessments into the regular banking activity. Increasingly more decision makers at national, European and international levels use the term "sustainable banking" and claim for immediate action in order to increase investment in longer-term, environmental and societal-friendly projects.

Although sustainable banking is considered to be an evolving concept lacking a formal definition, international authorities have arrived at a consensus: it encompasses all the investment decisions and financial flows channeled to support various economic sectors by also considering environmental, social and government (ESG) issues.

The European Commission (2019) further explains each aspect to be considered by banks: i) environmental concerns are related to climate change mitigation and reducing waste; ii) social concerns refer to issues of gender and wage inequality, financial inclusion, labor relations, investment in human capital and local communities; and iii) governance is related to management structures, employee relations, setting the tone at the top for ensuring the inclusion of social and environmental considerations in the business strategy, risk management framework, banking products and services' features.

Another definition of sustainable banking was issued by the European Banking Authority (2019) and gravitates around its goals, namely:

- integrating ESG criteria into the classical financial services;
- supporting sound, sustainable, smart and inclusive economic growth;
- increasing financial intermediaries' awareness of the need to mitigate ESG risks by including them in the internal risk management framework.

The International Monetary Fund (2019) advocates for a simpler general definition which is related to the incorporation of ESG principles into banking investment strategies and business decisions for triggering positive public externalities and economic development.

A more detailed definition of sustainable banking has been provided by the International Finance Corporation (2017), which ascertains that various definitions may exist across regions and communities depending on local practices and preferences. Therefore, the term may include three components, namely:

1   environmental and social risk management related to investment and lending activities;
2   green loan origination and investment to economic sectors and projects which bring positive environmental and societal (E&S) impact;
3   management of banks' own E&S footprints through energy and power saving and corporate social responsibility initiatives.

The first two components are core ones; however, they are differently weighted across banks and countries. The same report delineates between sustainable and green banking, arguing that the term "green banking" is often approached from the environmental perspective. Therefore, sustainable banking is a more comprehensive concept that embeds the green banking one.

The reasons underlying banks' decisions to commit to sustainable business practices are diverse, ranging from their will to design and provide innovative green financial products in order to gain more customers and differentiate from their primary competitors, to better managing ESG risks as claimed by international authorities, to avoid being exposed to reputational risks due to noncompliance with sustainability frameworks and actions taken at the European level, to be associated with the label of ethical financial institution or simply to follow the trend and keep pace with competitors so as not to lose their competitive advantage (herding behavior). Ethical financial behavior encompasses a bank's sustainable behavior, as ethics means doing the right thing in the benefit of a larger community while being motivated by good, rational reasons and taking appropriate actions in this regard (Bhala, 2019).

Recent frameworks and action plans adopted at European and international level in the field of economic responsibility and climate change are placing banks at the forefront of this revolution, with widespread sustainable banking practices becoming a necessity and not a voluntary option. Consequently, banks have to profoundly adjust their governance and business models, as well as their mission, strategy and range of financial services provided to customers. The traditional short-term thinking on improving profitability, maximizing shareholder value and customer satisfaction has already become narrow and simplistic. It has to be complemented with a forward-looking perspective related to the positive, long-lasting impact the bank exerts on society and the environment through its own internal activities and indirectly through the businesses it had financed (Boitan, 2014).

In this regard, four complementary dimensions of a sustainable business model have been identified, namely: i) *financial sustainability* – the bank has to stay financially viable, resilient to shocks and well capitalized in order to ensure the going concern of the business; ii) *economic sustainability* – the projects/counterparties founded by the bank have to exert a positive contribution to the national economy; iii) *environmental sustainability* – diminishing waste and preservation of natural resources; and iv) *social sustainability* – the projects financed by the bank have to address poverty reduction, improving the quality of life and welfare (International Finance Corporation, 2005).

To help banks redefine their business strategy into a sustainable one, BankTrack (2006) has issued several guidelines. A bank should start with an in-depth assessment of its current strategy, capabilities and projection of future achievements and ambitions. Within this framework, which is bank-specific, will be included those social and/or environmental aspects the bank is willing to assign top priority to. Then the bank has to choose those sectors of activity, regions/countries and customers it will focus on, so that banking activity makes a real contribution in fostering sustainability. The next step is to design new financial products or improve the existing range in order to better satisfy the financing needs of targeted communities.

Therefore, once the bank commits itself to a sustainable path, it has to define internal ethical policies to be followed by top management and employees, but also some minimum standards to be met by each retail and corporate customer, before entering into financial relations with the bank. This steady and transparent behavior, showing no forbearance to harmful environmental or climate businesses or projects, will ensure the bank's success in achieving its sustainable mission.

Box 4.1 Examples of sustainable business models implemented by banks

BNP Paribas employs the most complete strategy by reconciling four different financial models in order to combine economic performance with a positive social and environmental impact: i) socially responsible investing, which integrates sustainability into management; ii) green finance, which is related to ecological bonds; iii) social finance, which targets those investments into funds to the benefit of local communities; and iv) social business, which aims at reinvesting profits into social projects.

Triodos Bank defines itself as a values-based bank which relies on money entrusted by depositors in order to bring positive social, environmental and cultural impact. Its business model makes prevailing use of deposits attracted, rather than financial resources borrowed from other banks, as the main source for its lending activity.

Nordea, one of the world's leading sustainable banks, in 2019 integrated sustainability into all business activities and products within core areas of investment, financing and advisory services, as well as in internal operations. It launched a series of green financial products suitable to depositors, borrowers and investors in order to enable its customers to make conscious sustainable choices.

ING implements a strategy centered on identifying and financing business opportunities meant to facilitate society's shift to sustainability. All financing and investment policies apply rigorous and strict social, ethical and environmental screening criteria.

Intesa Sanpaolo defined its business model as being at the service of the real economy, through a strong commitment to supporting families and business in the communities it operates and development of a circular economy.

Commerzbank is aware of its economic, ecologic and social responsibility. In order to better coordinate the activities and business lines linked to sustainable finance, the bank has created a new internal structure called the Sustainable Finance Committee. Its aim is to facilitate the exchange of information on current issues regarding sustainable finance among middle and top management and to support the development or improvement of sustainable banking products.

The online bank mBank from Poland follows a business strategy articulated around five core goals: respect for customers' values and needs, acting as a responsible lender, investing in human resources for building an exceptional team, limiting the impact on the environment and improvement of the managerial approach.

The Co-operative Bank places at the core of its business the ethical banking conduct, with a strong emphasis on environmental safeguards. The bank finances only customers and projects that support a healthy environment and tackle climate change and actively seeks to minimize its environment footprint (zero landfill waste, use of renewable energy sources, reducing job-related travels).

CaixaBank implements a socially responsible banking model based on sustainable profitability; proximity and social commitment to its stakeholders; high ethical conduct; and governance, quality and trust.

A bank's board should implement a top-down approach in spreading sustainability concerns across the banking business. The entire range of banking products has to be assessed from the standpoint of their positive or negative impact on social and environmental sustainability. According to BankTrack (2006), this has to consider retail banking (saving accounts, credits, mortgages), corporate banking (company loans, trade finance), investment banking (stock and bond issuance and trading, project finance, stock analysis, Merger & Aquisitions and other corporate advising), asset management, private banking and any other forms of financial services.

Professionals in the banking system as well as authorities agree that banks have to reconfigure their main activity – lending – by designing new screening processes in order to identify economically sound projects and businesses which at the same time are environmentally friendly and socially responsible. In addition, apart from monitoring and assessing the credit risk associated with the loan portfolio, banks have to consider the E&S risks caused by the implementation of the projects founded by the bank. Failure to address E&S risks may subsequently trigger financial losses for the bank and reputational risk.

The International Institute for Sustainable Development (2013) has noticed that banks have already implemented sustainability issues into their business model. More specifically, the internal daily operations were adjusted to incorporate environmental initiatives (such as recycling programs, energy and resource saving programs, solar-powered bank branches and automated teller machines [ATMs]) and socially responsible initiatives (such as sponsorships for educational or cultural events, improved human resource practices, charitable donations). A second category of measures which is expected to generate a larger-scale impact resides in the integration of sustainability into the bank's core business and decision-making process, with an emphasis on lending and investment and the development of new green products.

A report published by World Finance (2019) observes that banks across the globe are keeping pace with decision-makers' financial agendas by diversifying their range of financial products to include socially responsible investment products. It warns that failing to invest in sustainable products and services may be detrimental for a bank's credibility and profitable going concern.

Recently, several large-scale banks have designed and launched various innovative financial products, such as:

- *green loans* – represent loans specifically designed for environmental businesses, such as wind farms, solar energy, geothermal energy, waste management, water treatment and supply, circular economy solutions, organic food and agriculture. As the International Finance Corporation

(2017) outlines, there may be differences between country-specific green loan categories, as they used to be defined based on national strategic priorities, economically important and high-impact sectors and local E&S challenges. Also, there is the perception that green lending is more costly than traditional lending, as screening processes require increased due diligence and stricter selection of projects/borrowers to be founded. Most ethical lending policies adopted by banks envisage the decrease of the financing provided to projects related to fossil fuels and coal power generation.

- *green or sustainable mortgages* (energy-efficient mortgages) –loans granted to household customers in order to create incentives for living in climate-smart, environmentally friendly homes (green buildings). For instance, retail customers may be charged considerably lower interest rates if they purchase new energy-efficient homes and/or invest in energy-efficient appliances. Other banks are covering through this type of loan the cost of switching a house from conventional to green power.
- *green commercial building loans* – a loan created for financing green commercial buildings, characterized by lower energy consumption, reduced waste and less pollution than traditional buildings (United Nations Environment Program – Financial Initiative, 2007).
- *green car loans* – loans provided at a below-market interest rate to support the purchase of electric vehicles or vehicles with high fuel efficiency.
- *sustainability improvement loan* – represents a loan provided to companies at a lower interest rate, conditioned by the fact that the corporate customer has to improve its sustainability performance, including climate performance (it is granted by ING Bank).
- *loans for high-social-impact activities* – these are currently issued by the Intesa Sanpaolo Group and consist of loans meant to support the development of entrepreneurship and employment opportunities or to assist people in difficulty (through different forms such as microcredit; anti-wear loans; loans for third-sector associations; funding to support people affected by catastrophic events). Usually, these loans are unsecured and therefore expose the lender to higher risks. In adopting the lending decision, an analysis is always performed that reconciles the economic viability with the social impact to be brought for local communities. Social impact lending is a valuable tool for mitigating financial exclusion by alleviating the financial challenges faced by small companies and start-ups, charities or individuals. It favors those businesses and projects that have the potential for wider community benefit or social impact.
- *basic current account* – this is dedicated to senior customers to increase their financial inclusion and has two components. The banking component provides simplicity, transparency, low management costs and free

transactions, while the non-banking component brings advantages in terms of health, wellness and recreation. More specifically, it provides protection against unforeseen events, access to medical and healthcare services, welfare and tax assistance services at subsidized rates or even free opportunities to purchase recreational products and services at preferential prices (developed by Intesa Sanpaolo Group).

- *green savings account* – it is a savings account which aims to attract people concerned with using their money consciously by giving them a sustainable value. The deposited money will be used by the bank in order to finance new, high-impact sectors.
- *green credit cards* – the bank will donate a percentage of a cardholder's spending to environmental causes or to a specific non-profit organization. Barclays has issued a specific green card, called a climate credit card, which provides discounts and low borrowing rates to its users, but only when they purchase environmentally friendly products and services (energy-efficient appliances, public transportation passes). In addition, half of the after-tax profits generated by the use of this card will be used by the bank to finance carbon emission reduction projects worldwide.
- *green certificates of deposit* – represent certificates of deposit issued by a bank with the aim to finance a portfolio of eligible green assets, while at the same time providing a sustainable debt instrument for interested investors.
- *green bonds* – bonds issued by banks with the same contractual characteristics as classical bonds but with a specific goal: all proceeds obtained by the bank have to be invested in projects with clear environmental benefits (renewable energy, energy efficiency, sustainable waste management, biodiversity, innovative and clean transportation). The first green bonds were issued by the European Investment Bank and the World Bank in 2007. Since the signing of the Paris Agreement in 2016, these bonds achieved momentum as financial institutions become fully aware of their pivotal role in the transition to a low-carbon economy and shifted from trading stocks to sustainable bonds.
- *forest bonds* – represent green bonds with a specialized focus and extended maturity. An example of a forest bond is the one meant to fund large-scale reforestation in Panama, being issued with a 25-year maturity.
- *social bonds* – represent bonds whose proceeds will be exclusively channeled to eligible social projects. Similar to green bonds, they are regulated debt instruments as any other listed fixed-income securities.
- *sustainability bonds* – represent bonds whose proceeds are used to finance or refinance a combination of environmental and social projects. A report published by IMF (2019) highlights that green bonds

are leading the global issuance of sustainability-linked bonds and that Europe is the largest issuer of green bonds, followed by the Asia-Pacific region and North America. Méndez-Suárez et al. (2020) argue that the various types of sustainability bonds provide an opportunity to reach the Sustainable Development Goals, improve the sustainable investment portfolios and strengthen a bank's corporate social responsibility policy and its corporate reputation.

- *green mortgage-backed securities* – represent fixed-income debt instruments originating through the securitization of green mortgage loans.
- *green fiscal funds* – represent funds that are launched and administered by the bank. People purchasing shares in such a green fund are exempted from paying capital gains tax and receive a discount on income tax, but they also receive a lower interest rate for their investment in the fund. The bank relies on amounts attracted through green funds to further provide green loans at a lower cost to finance environmental projects (United Nations Environment Program – Financial Initiative, 2007).
- *green digital finance* – represents the use of digital technologies and innovations for achieving objectives related to sustainable development. A Chinese bank is pioneering in this regard, as it tried to convince people to adopt a greener lifestyle (engage in low-carbon activities, paying utility bills online, walking or cycling instead of driving, etc.). The bank provided its users "green energy" points: when a certain level of energy points was reached, the bank and its philanthropic partners would plant a tree in the desert.

To sum up, the range of green banking products developed so far is diverse, and banks are claiming that the good revenues obtained in the past few years is certainly due to the implementation of these clean financing activities. As the International Finance Corporation (2017) explains, in order for a bank to have access to the sustainability playing field, it has to overcome five common barriers, namely motivation barriers, information/knowledge barriers, technical barriers, financial barriers and customer awareness barriers.

Additional obstacles to sustainable banking are related to the uncertain bankability of projects requesting financing, to low transparency in tracking sustainable capital flows and to the lack of a harmonized framework for matching green investment supply and demand (Mendez, Houghton, 2020).

From the customers' standpoint, the reasons for choosing a green, social impact financial product are extremely diverse. Some may be attracted by the features of these products (returns, financial and additional non-financial benefits); others are more interested in providing their savings an ethical destination, while other individuals perceive green financial products as a long-lasting tool for protecting human rights and the environment and mitigating global warming. There is also a group represented by institutional investors which is continuously searching for alternative placements and exhibits increased interest in diversifying their portfolios with socially responsible or ethical investments in green bonds and securities.

Regulatory initiatives and international guidelines and principles for achieving responsible banking activity

In the following the regulatory initiatives, action plans and agreements launched by the European Commission, the European Banking Federation, the European Central Bank and the European Banking Authority are summarized in order to support a smooth transition of the financial sector towards sustainable finance and growth.

At the international level, two influential agreements concluded in 2015 have prepared the way towards building a global financial system that supports sustainable growth and climate-resilient development. They are represented by the United Nations 2030 Agenda for Sustainable Development (which defines 17 Sustainable Development Goals covering the three dimensions of sustainable development: economic, social and environmental) and the Paris Climate Agreement, the first global climate change agreement aiming at achieving a state of climate neutrality before the end of the century.

Subsequent to these global initiatives, the European Commission constituted in late 2016 a high-level expert group on sustainable finance, whose recommendations were included in the action plan on sustainable finance adopted by the commission in March 2018. The novel and comprehensive key actions, meant to strengthen the connection between financial systems and sustainability, comprise:

- the development of a unique and detailed classification of sustainable activities performed by all market players in the financial system, called EU taxonomy;
- setting up EU labels to facilitate the identification of green financial products;
- establishing clear institutional investors' duties regarding sustainability;
- fostering the transparent disclosure of information on companies' ESG policies;
- introducing the term "green supporting factor" in the EU prudential rules by incorporating environmental-related risks into banks' risk management policies and potential recalibration of capital requirements, without compromising financial stability.

In 2019 the EC's expert group released two reports, one related to establishing an EU green bond standard and another focusing on the EU taxonomy.

In parallel, in June 2019 the European Banking Federation and the United Nations Environment Programme Finance Initiative (UNEP FI) launched a joint project aiming at assessing the extent to which the EU taxonomy on sustainable activities could be applied to core banking products and the development of best practice guidelines in this regard. The first recommendations are expected to be delivered at the end of 2020.

Complementary to these approaches, in December 2017, 54 central banks and supervisory authorities across the world established the Network for Greening the Financial System. It is a global network aimed at integrating sustainability-related risks into the financial supervision process and portfolio management, knowledge sharing and advocating for a financial system that supports low-carbon economic growth.

The European Central Bank (2019) also is concerned about the impact climate-related risks may exert on financial stability. A preliminary analysis has indicated that climate risks in the financial system have the potential to become systemic in the euro area unless this specific risk is not priced correctly.

Apart from establishing road maps and action plans and issuing regulations for better connecting financial systems with Sustainable Development Goals, the International Monetary Fund (2019) outlines that decision makers and central banks have to step in and provide intellectual leadership in the process of ESG risk assessment. Multilateral cooperation may support further development in the supervisory process of ESG risks, with the recommendation being that these specific risks be included into financial stability monitoring and into micro-supervision tools (such as stress testing).

De Nederlandsche Bank (2017) emphasizes several issues of interest from a prudential supervision standpoint. Three risks could possibly emerge as a consequence of green finance development, namely i) green bubbles; ii) reputational risks due to green-washing (ambiguous definition of green investments); and iii) relaxed regulatory requirements meant to promote climate-related investments. It is strongly advised that supervisory rules, especially those related to capital requirements, shouldn't be lowered in order to promote further involvement of financial institutions in financing sustainable projects. The capital requirements have to adequately reflect all the risks, including the ESG ones, and help the financial institution withstand potential financial and macroeconomic shocks or crises; therefore, they should not be lowered to achieve social or environmental objectives.

A recent position paper issued by the German association of banks (German Public Banks, 2019) welcomes the European Commission's initiative in setting up a common classification for sustainable financial products, which is expected to enhance transparency and address uncertainty for investors, as well as to contribute to financial stability. When developing this taxonomy, the focus should be on environmental aspects. With regard to the widely debated issue of regulatory capital relief for the provision of green loans, it is suggested that it should be accepted, provided that those loans expose the banks to significantly lower risks than regular loans.

This view is shared also by the European Banking Federation (2019), which suggests that the capital adequacy issue should be analyzed at the European level, being the task of the European Banking Authority to employ forward-looking approaches in order to conclude whether preferential capital treatment should be given for certain sustainable assets that show a lower financial risk.

To sum up, there is global consensus among decision makers, central banks and supervisory authorities that ESG aspects, especially environmental-related ones, cannot be neglected anymore from the standpoint of both financial systems' soundness and sustainable development. However, a widely agreed taxonomy discriminating between green and brown activities, as well as a monitoring framework and adequate regulations, are still in progress.

In addition to the official developments already mentioned, the financial industry is witnessing momentum in voluntarily involving itself in various sustainable best practice initiatives. They take the form of international frameworks, guidelines, standards and principles for responsible banking activity issued by various international organizations, with the fundamental purpose of assisting banks in screening and financing socially and environmentally sustainable economic activities.

To examine the openness of contemporaneous banking systems across Europe for joining various sustainability standards and committing to fulfill specific transparency and responsibility criteria, a cross-country qualitative analysis was performed. More specifically, the membership of each main international sustainability framework/standard was analyzed to uncover which European banks have become signatories. Then the results were mapped to gain a visual representation of the most widespread sustainability standards across Europe, but also of countries whose banks have simultaneously joined several standards.

The UNEP FI was developed in 1992, having as its fundamental aim the increase of the financial industry's awareness of the environmental agenda. The statement of commitment focuses on three main sustainability goals: sustainable development, sustainability management and public awareness and communication. Banks that are members of this framework belong to 21 European countries. The visual plot of UNEP FI spread across the region is shown in Figure 4.1.

The United Nations Global Compact was launched in 2000. Its signatory members have to reshape their strategy by doing business responsibly, in

Figure 4.1 The spread of the UNEP FI framework

Source: Author, based on membership information from www.unepfi.org/members

compliance with 10 principles for human rights, labor, environment and anti-corruption. Figure 4.2 shows 17 European countries whose banks have voluntarily joined this sustainability framework.

The Equator Principles were launched in 2003 and have undergone successive stages of updates. The principles represent a risk management framework which is adopted by financial institutions in order to facilitate the identification, assessing and managing of environmental and social risks generated by financed projects. They aim at providing a minimum,

Figure 4.2 The spread of the UN Global Compact framework

Source: Author, based on membership information from www.unglobalcompact.org/what-is-gc/participants

harmonized standard for due diligence and risk monitoring, which are at the core of responsible risk decision-making. The principles target four financial products, namely i) project finance advisory services; ii) project finance; iii) project-related corporate loans; and iv) bridge loans. Figure 4.3 shows that, so far, 12 European countries have chosen to implement these principles in their business conduct.

Figure 4.3 The spread of the Equator Principles

Source: Author, based on membership information from https://equator-principles.com/members-reporting/

The Global Alliance for Banking on Values (GABV) was launched in 2009 and represents a network of banks committed to achieving economic, social and environmental sustainability while performing their regular business. GABV is proponent of a triple bottom-line approach (people, planet and prosperity), with banks fully aware of the local communities' needs and

Figure 4.4 The spread of the Global Alliance for Banking on Values

Source: Author, based on membership information from www.gabv.org/the-community/members/banks

involved in increasing financial inclusion. Figure 4.4 indicates that banks from 11 European countries have joined this network.

The RE100 initiative (renewable energy 100%) was launched in 2014 through a partnership represented by the Climate Group and Carbon Disclosure Project. The signatories of RE100 publicly commit to sourcing 100% of their electricity consumption from renewable energy sources by a specified year (2050 at the latest). Its purpose is to accelerate the transition towards zero carbon grids at a global scale. Banks from eight European countries have committed so far to the requirements of this framework (Figure 4.5).

In 2017 the International Capital Market Association launched a series of three principles, called the Green Bond Principles, the Social Bond Principles and the Sustainability Bond Guidelines, with the aim of creating a common

Figure 4.5 The spread of the RE100 initiative

Source: Author, based on membership information from http://there100.org/companies

framework at a global level for the issuance of green, social and sustainability bonds. These voluntary guidelines promote increased transparency, disclosure and integrity in the issuance and use of proceeds generated by green/social/sustainability bonds. Although recently established, these guidelines have attracted banks from 16 European countries to their membership (see Figure 4.6).

The Principles for Positive Impact Finance were released in 2017, being launched at the initiative of the UNEP FI. They are rooted on a unique theory of impact, which aims at analyzing the impact exerted by financial

Figure 4.6 The spread of green, social and sustainability principles

Source: Author, based on membership information from www.icmagroup.org/green-social-and-sustainability-bonds/membership/

intermediaries' business on Sustainable Development Goals. By committing to these principles, it is envisaged that the emergence of new, impact-based business models will trigger a positive contribution to one or more of the three pillars of sustainable development (economic, environmental and social). Figure 4.7 shows the eight European countries whose banks have voluntarily joined these principles.

Figure 4.7 The spread of Principles for Positive Impact Finance

Source: Author, based on membership information from www.unepfi.org/positive-impact/members-2/

The newest initiative launched in late 2019 by UNEP FI is represented by the UN Principles for Responsible Banking. They aim at becoming the most significant partnership between the global banking industry and UNEP FI. The six principles are meant to foster the banking system's involvement in achieving the Sustainable Development Goals and the Paris Climate Agreement by helping any interested bank in aligning its business strategy with societal goals. These principles have succeeded in attracting banks from 12 European countries (see Figure 4.8).

Figure 4.8 The spread of the UN Principles for Responsible Banking

Source: Author, based on membership information from www.unepfi.org/banking/bankingprinciples/signatories/

By reconciling all the information previously described, it seems that European banks have actively stepped into the sustainability field by adopting several voluntary international guidelines and principles. The various frameworks are meant to complement themselves and are continuously updated to keep pace with the development of the financial industry and the environmental and societal challenges.

One finding is that the UNEP FI framework has attracted the highest number of countries whose banks became signatories, followed by the UN Global Compact and the green, social and sustainability principles. Therefore, at the European level these standards are the most successful and widespread. At the opposite lie the RE100 and the Principles for Positive Impact Finance, which have attracted the lowest number of member banks, located in only eight European countries.

Second, it can be noticed that some European countries (France, Germany, Italy, Norway and the UK) appear in the membership of all eight sustainability frameworks, whereas three countries (Denmark, the Netherlands, Spain) are present, through their banks, in the membership of seven out of eight frameworks. There are also European countries (Albania, Czech Republic, Estonia, Latvia, Lithuania, Slovenia) witnessing no single bank in the membership of a sustainable framework.

Consequently, banking systems in some countries exhibit more sustainable features in terms of reshaping business strategies, reconfiguring the characteristics of products and services, improved risk management framework to comply with ESG risks, increased transparency and due diligence, while others are still rooted in the traditional conduct of the banking business.

Acknowledgement: The documentation process for the elaboration of this book was partially supported by a mobility grant of the Romanian Ministry of Research and Innovation, CNCS – UEFISCDI, project number PN-III-P1–1.1-MC-2019–0990, within PNCDI III.

References

BankTrack. (2006). *The do's and don'ts of sustainable banking – a banktrack manual*, www.banktrack.org/download/the_dos_and_donts_of_sustainable_banking/061129_the_dos_and_donts_of_sustainable_banking_bt_manual.pdf.

Bhala, T.K. (2019). The philosophical foundations of financial ethics, in *Research handbook on law and ethics in banking and finance*, Edward Elgar Publishing Limited, https://doi.org/10.4337/9781784716547.

Boitan, I. (2014). *The social responsibility stream in banking activity. A European assessment*, Conference proceedings SGEM: Political sciences, law, finance, economics and tourism, published by STEF92 Technology Ltd., Sofia, Bulgaria, ISSN 2367–5659, pp. 793–800.

De Nederlandsche Bank (DNB). (2017). *Waterproof? An exploration of climate-related risks for the Dutch financial sector*. De Nederlandsche Bank, Amsterdam.

EU Technical Expert Group on Sustainable Finance. (2019). *Report on EU green bond standard*, https://www.buildup.eu/en/node/57836.

EU Technical Expert Group on Sustainable Finance. (2019). *Taxonomy: Technical report*, https://ec.europa.eu/energy/sites/ener/files/documents/20170125_- _technical_report_on_euco_scenarios_primes_corrected.pdf.

European Banking Authority. (2019). *EBA action plan on sustainable finance*. December, https://eba.europa.eu/sites/default/documents/files/document_library/EBA%20Action%20plan%20on%20sustainable%20finance.pdf.

European Banking Federation. (2019). *Encouraging and rewarding sustainability*, www.ebf.eu/wp-content/uploads/2019/12/ENCOURAGING-AND-REWARDING-SUSTAINABILITY-Accelerating-sustainable-finance-in-the-banking-sector.pdf.

European Central Bank. (2019). Climate change and financial stability. *Financial Stability Review*, ECB, May.

European Commission. (2019). *Green finance – overview*, https://ec.europa.eu/info/business-economy-euro/banking-and-finance/green-finance_en#ipsf.

German Public Banks. (2019). *Current positions on the regulation of banks and the financial markets*. Bundesverband Öffentlicher Banken Deutschlands – VÖB, October.

International Finance Corporation. (2005). *Choices matter – sustainability report*, www.ifc.org/wps/wcm/connect/1fbec523-e2f3-425a-9e2d-8ddff152def7/SR_Summary_English.pdf?MOD=AJPERES&CACHEID=ROOTWORKSPACE-1fbec523-e2f3-425a-9e2d-8ddff152def7-jqeIZyc.

International Finance Corporation. (2017). *Greening the banking system – experiences from the sustainable banking network* (SBN), www.ifc.org/wps/wcm/connect/5962a2da-1f59-4140-a091-12bb7acef40f/SBN_PAPER_G20_02102017.pdf?MOD=AJPERES&CVID=lHehxyG.

International Institute for Sustainable Development. (2013). *Sustainable banking*, www.iisd.org/business/banking/sus_banking.aspx.

International Monetary Fund. (2019). *Global financial stability report*, Washington, DC, October.

Mendez, A., Houghton, D.P. (2020). Sustainable banking: The role of multilateral development banks as norm entrepreneurs. *Sustainability*, 12(3), 972, https://doi.org/10.3390/su12030972.

Méndez-Suárez, M., Monfort, A., Gallardo, F. (2020). Sustainable banking: New forms of investing under the umbrella of the 2030 agenda. *Sustainability*, 12, 2096, https://doi.org/10.3390/su12052096.

United Nations. (2015). *Transforming our world: The 2030 agenda for sustainable development*, https://sustainabledevelopment.un.org/post2015/transformingourworld/publication.

United Nations Environment Program – Financial Initiative. (2007). *Green Financial Products and Services*. Current Trends and Future Opportunities in North America, www.unepfi.org/fileadmin/documents/greenprods_01.pdf.

World Finance. (2019). *Sustainable banks*. www.worldfinance.com/banking-guide-2019/sustainable-banks.

Chapter 5

Designing sustainable financial instruments

Malgorzata Tarczynska-Luniewska
and Sebastian Majewski

Introduction

The basis for the creation of sustainable financial instruments should be found in sustainable finance. In particular, the problem of developing this category of tools related to sustainable investments, where appropriate tools are needed to make investment decisions. The development of the capital market, including the issue of sustainability investment, has forced the need to construct financial instruments. Designing sustainable financial instruments is a process in which the essential role is played by:

1 the foundations for the creation of this type of instrument,
2 methods, which provide a methodological framework for their design,
3 information, which includes quantitative and qualitative data enabling the issues of measurement, analysis, diagnosis or forecasts as part of investment decisions made with the use of sustainable financial instruments.

Sustainable financial instruments created in response to the needs of market development can be associated with actions and decisions taken as part of socially responsible investing. Sustainable financial instruments designed in response to the needs of market development can be associated with activities and decisions taken as a part of socially responsible investing. Such a situation refers to the philosophy of investing, where traditional criteria are extended and complemented by non-financial factors involving social and environmental aspects [Dasgupta, 2007]. The idea of ethically based investment is not a new concept. The very first fundamentals of this kind of investment were canons of faith and determinants of moral behaviour. Its dimension is similar in Judaism, Christianism and Islam, and it treats usury, gambling, pornography or pork trade on the same level of prohibition [Louche & others, 2012; Abraham, 2010]. Therefore, socially responsible investment was largely based on the pillars of individual religions or religious organizations.

Socially responsible investment (SRI) means part of a financial analysis is focused on social, environmental and ethical issues during processes of selecting, managing and liquidating investments [Mansley, 2000]. SRI includes investments that integrate social, ethical, environmental and corporate governance into the investment process [Sandberg & others, 2009]. Despite fact that the first socially responsible fund was founded in the United Kingdom (Stewardship Fund in 1984), the United States is the forerunner of responsible investment. The main reason for developing such kind of investment was not ethical behaviours connected to religion, but social protests against the wars, environment pollution and lack racial and gender equality. The ever more important subsequent social protests expanded the list of reasons for the necessity of developing SRI (the next causes were against Apartheid in South Africa and atomic energy technology after some accidents in nuclear power plants).

The idea of stakeholders such as R.E. Freeman [Freeman, 1984] is a key point of most of the concepts developed over the years. Freeman classifies shareholders into two groups: shareholders interested mainly in the profit from the company's activities and stakeholders interested in the company's operations in the broad sense. This theory states that the activity of a company in a given area obliges management to consider in its strategies in terms of relations with the local community, environmental protection and other non-economic objectives. However, there is no empirical evidence for the influence of different groups of shareholders on the decision of implementing environmental, social and governance (ESG) issues in companies.

The SRI concepts ceased to be a negative selection and began to focus on improving the investment process, so that in addition to financial goals, it included good practices in social and environmental cooperation at the end of the 1980s [Krosinsky et al., 2011]. It is worth noting that SRI has many synonyms [Overview of European Sustainable Finance Labels, 2019]. This fact has a strong impact on designing and the name of financial sustainability instruments (i.e. in Poland the RESPECT index refers to socially responsible and sustainably managed companies listed on the Warsaw Stock Exchange, and it can be used as the sustainable financial instrument).

Above all, both sustainable financial instruments and the SRI are voluntary and depend on the goodwill of the investor. While it has been observed that non-governmental organizations (NGOs) are more committed to supporting and promoting companies with high ethical standards, there are no prohibitions from investing in companies with low ethical standards. NGOs often lobby for companies engaged in activities for sustainable development, which has a result of the creation of proper regulations, both national and supranational (including codes and standards of good practice), in this scope. One of the important factors shaping the policy of sustainable functioning of enterprises is an initiative of the United Nations called the

Principles of Responsible Investment (PRI). The most important of them are [Majoch & others, 2017]:

- the application of ESG standards to analyses, supporting investment processes and the investment decision-making process itself,
- ownership activity understood as a strategy of involvement and voting and integration of ESG issues into the company's policies and activities,
- an attitude of transparency with respect to entities affiliated with the company on ESG issues,
- lobbying for the implementation of ESG principles in the investment sector,
- increasing the effectiveness of the implementation of SRI principles by developing cooperation between entities,
- reporting on activities and progress in implementing SRI principles.

Factors with a significant impact on the possibilities of developing sustainable financial instruments are also the increased interest of the society and the level of economic development of the country. Unfortunately, this type of investment instrument is characteristic of societies with a very high economic awareness, which is a result of high economic development.

Sustainable financial instruments

Among the institutions offering financial products to individual investors one can distinguish financial institutions, government and self-government institutions, foundations, non-governmental institutions and universities. They offer individual investors the possibility to use many financial products and services that correspond to the sustainable development policy. These are, among others [Dikau & Volz, 2018]:

- banking deposits (for example, earmarked deposits financing the protection of some endangered species – banks transfer a small amount from each deposit to save endangered animals),
- credits (for example, ecological mortgage with a lower-than-ordinary rate in case of using ecological sources of energy – solar collectors, recuperators or heat pumps),
- credit or debit cards (analogical situation as in the case of deposits – a small part of banking fees or commissions for transactions made is transferred on the environment protection),
- structured products – created on the basis of equities and bonds of ecological issuers or socially responsible stock exchange indexes,
- bonds (for example, green bonds, where investors have opportunities to financially support ethical or ecological projects connected with protecting the natural environment)

- investment or pension funds supporting sustainable development
- responsible stock exchange indexes.

Green investments relate to the financial activities of enterprises targeted at reducing greenhouse gas emissions and air pollution without significantly reducing the production and consumption of non-energy goods. It takes into account, in particular, emissions of greenhouse gases (especially carbon dioxide) and pollutants (such as sulphur dioxide and nitrogen oxide) leading to global warming, smog and acid rain and that have a negative impact on health. Of course, other environmental objectives may also be considered [Stern & Stern, 2007]. These may be, e.g. reducing dependence on fossil fuels, avoiding depletion of resources, preventing water and soil damage, reducing waste and preserving biodiversity. Eurostat adopted a broad approach in 2009 defining environmental expenditure as the acquisition of technologies, goods and services focusing on reducing the degradation and depletion of natural resources. Environmental strategies of companies are often focused on measures aimed at introducing low-energy supply chains, increasing energy efficiency and reducing the use of energy derived from coal combustion. The other variant may be the supporting of investment in renewable energy. Here investments in nuclear energy could come in, but it is not considered for a number of reasons, like radioactive waste is often generated, expenditures on nuclear energy are often high and financed by the public sector and the development of nuclear energy is more dependent on technological progress [Eyraud et al., 2013].

Taking into account all possibilities for obtaining money to be invested in the sustainable economy, it could be underlined that stock exchange instruments are the most effective. One of the most popular stock exchange indexes dedicated to clean energy is the NEX. The WilderHill New Energy Global Innovation Index (NEX) was constructed in 2006 and contains companies mostly outside the United States using innovative technologies and services focused on the generation and use of clean energy, lower CO_2 renewables, conservation and efficiency. Figure 5.1 presents the quotations for NEX.

The NEX is based on modern portfolio theory and the rule of intelligent indexing. Very important also is an analysis-based selection of stocks and sector weightings, which are reviewed according to both qualitative and quantitative methodology. The index structure is changed quarterly, taking into account the sustainable balancing rule – the goal of the index isn't to beat the market or to invest in undervalued stocks. NEX mainly consists of companies in wind, solar, biomass and biofuels; small-scale hydro; geothermal; marine; and other relevant renewable energy businesses. In its structure are also noted enterprises targeting step-change improvements in generation, distribution and storage of energy, as well as conservation, efficiency, materials and in the emerging hydrogen and fuel cell sectors and associated services.

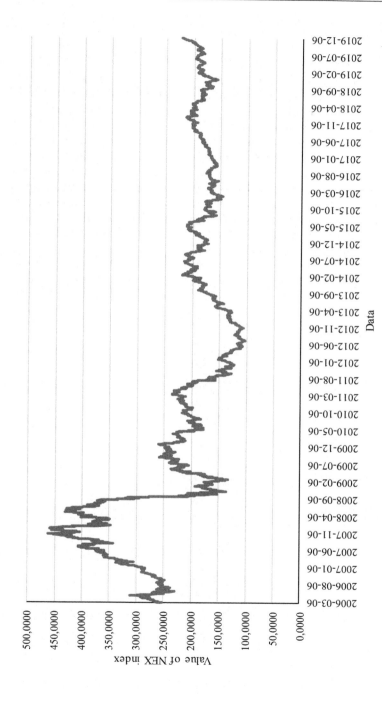

Figure 5.1 Quotations of NEX index

Source: https://nexindex.com/about_nex.php

An investor cannot directly invest in the index. This is possible thanks to index-tracking funds, which try to closely imitate NEX performance. The possibility could be done by investing, for example, in the Invesco Global Clean Energy ETF.

Analyzing the potential opportunities offered by the market, a steady increase in the number of investment funds offering their clients ethical financial instruments is observed. Generally, it can be assumed that SRI funds are divided as shown in Figure 5.2.

The three groups of funds: environmental, ethical and social, make it clear what ethical issues the group relies on. In the case of religious funds, the homogeneity of the group cannot be established. Therefore, they deserve special attention. The literature focuses here on three groups of religious funds: those based on Judaic [Wilson, 1997], Christian [Renneboog & Spaenjers, 2012] and Muslim ideology [Abraham, 2010]. In fact, the latter group is recently one of the fastest developing – which has created a scientific sub-discipline of Islamic finance.

The criteria and standards of ESG became more relevant also for investors interested in exchange trade funds (ETFs). They had a strong influence on EFT's market. The large jump was observed from 2009 to June 2019 – from

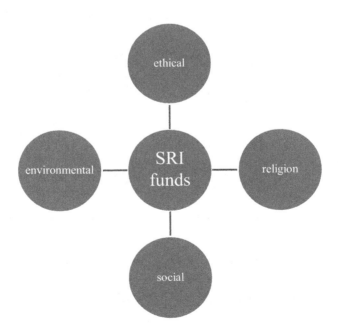

Figure 5.2 The structure of SRI funds

Source: Prepared on the basis of www.sriservices.co.uk/

39 (USD 5 billion of assets under management) to 221 funds (USD 25 billion). Such an explosion of interest in ESG ETFs was caused by two general factors [United Nations UNCTAD, 2019]. The first was governments enacting special preferential regulations that aim to foster more sustainable investment practices. Second, investors take into account the necessity of using investment vehicles based on sustainability criteria [Fulton & other, 2012]. The rising trend in ESG ETFs is also supported by some additional factors like:

- increased awareness of sustainable investing,
- there is a lack of evidence that sustainable investments are not profitable,
- increased awareness of sustainable investing of ESG EFTs (caused by increase awareness of sustainable investing),
- providers of indexes and funds adjust their offers by responding to the rising level of demand for sustainable investing and passive investing.

Sustainable business models

Aspects of the ESG may involve many factors that result in the construction of a business assessment model that can be used to monitor the situation of companies. For example, see an assessment of the MSCI index (Morgan Stanley Capital International) at www.msci.com/index-methodology presented in Table 5.1.

Table 5.1 ESG criteria

ESG criteria				
Environmental	Social			Corporate governance
	Society	Consumers	Employees	
management of natural environment protection	philanthropy	marketing and advertisement	employer– employee relations	sustainable development
climates changes	influence on local societies	quality and security of products/ services	safe working practice	corporate governance
CO_2 emissions	human rights	monopolistic practices	diversity in workplace	business ethics
renewable source of energy				political correctness

Source: Prepared on the basis of www.msci.com/index-methodology

A business model is a conceptual tool used to help understand in which way an enterprise does business. It is also used for analyses, comparisons and performance assessments, management, communication and innovation [Osterwalder & others, 2005].

Business models concern a way of defining competitive strategy. They are made, among other methods, through the design of the product or service, determining the level of costs to produce, how the firm differentiates itself from other firms by the value proposition and how the firm integrates its value chain with those of other firms in a value network [Rasmussen, 2007]. The literature indicates quite a large group of archetypes of sustainable business models presented in work of Bocken and others [Bocken & others, 2014]. The classification of archetypes is as follows:

1   Technological archetypes:

    1.1   Maximize material and energy efficiency (examples: low carbon manufacturing/solutions; lean manufacturing; additive manufacturing; dematerialization of products or packaging; increase functionality),

    1.2   Create value from waste (examples: circular economy; cradle-to-cradle; industrial symbiosis; reuse, recycle, re-manufacture; take back management; use excess capacity; sharing assets; extended producer responsibility),

    1.3   Substitute with renewables and natural processes (examples: move from non-renewable to renewable energy sources; solar and wind-power based energy innovations; zero emissions initiative; blue economy; biomimicry; the natural step; slow manufacturing; green chemistry),

2   Social archetypes:

    2.1   Deliver functionality rather than ownership (examples: product-oriented Product Service System [PSS] maintenance, extended warrantee; use oriented PSS rental, lease, shared; result-oriented PSS pay for use; Private Finance Initiative [PFI]; Design, Build, Finance, Operate [BFO]; Chemical Management Services [CMS]),

    2.2   Adopt stewardship role (examples: biodiversity protection; consumer care – promote consumer health and wellbeing; ethical trade; choice editing by retailers; radical transparency about environmental/societal impacts; resource stewardship),

    2.3   Encourage sufficiency (examples: consumer education models, communication and awareness; demand management; slow fashion; product longevity; premium branding/limited availability; frugal business; responsible product distribution/promotion),

3   Organizational archetypes:

    3.1   Repurpose for society/environment (examples: not for profit; hybrid businesses, a social enterprise; alternative ownership: cooperative, mutual, collectives; social and biodiversity regeneration initiatives; the base of pyramid solutions; localization; home-based, flexible working),

    3.2   Develop scale-up solutions (collaborative approaches: sourcing, production, lobbying; incubators and entrepreneur support models; licensing and franchising, open innovation platforms; crowed sourcing/funding; "patient/slow capital" collaborations).

The construction of new archetypes of business-making models has one major aim – to embed sustainability into business purpose and processes, increase the importance of innovations, accelerate their introduction and reduce risks of implementation in practice. Each type of model could be implemented separately, but there are not any restrictions to join different types of models to fulfil particular goals of enterprises. During the process of creating the new sustainable business model of functioning, the company archetypes here could be used as the reference point.

Detecting the problem of sustainable development caused the significant development of sustainable and responsible investment throughout the world. The American Foundation Social Investment Forum first measured the level of sustainable and responsible investment in the world in 1995. The value of the assets was USD 639 billion. In 2018 this size increased more than 18-fold and the annual growth rate equals 13.6% [Report on US Sustainable, Responsible and Impact Investing Trends, 2018].

One of the most important parts of investment made in the global financial sector is clean energy investment. The global value of clean investments increased fivefold from USD 63 billion to USD 329 billion in 11 years (Figure 5.3).

Both companies and individual investors could allocate financial sources in many classes of assets. The report made for Bloomberg indicates the following classes of assets [Mills & Byrne, 2016]:

- venture capital and private equity – all funds are raised for purposes for developing companies in the clean energy industry, according to Bloomberg New Energy Finance (BNEF) exposure ratings,
- public markets – stocks and bonds offered on primary and secondary stock exchange markets issued by companies that are primarily involved in the clean energy industry according to BNEF exposure ratings,
- asset finance – new-build financing of renewable energy–generating projects,

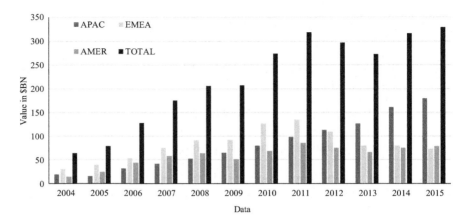

Figure 5.3 Clean energy investment in billions USD by sectors (APAC – Asia or Pacific; EMEA – Europe, Middle East and Africa; AMER – North and South America)

Source: Authors' study on the basis of data reported by Mills and Byrne [Mills & Byrne, 2016].

- small-scale solar – investing in house production of solar energy,
- R&D – investing in research and development activity.

Generally, providing business-respecting sustainable goals requires entrepreneurs to transform the traditional business model into an approach respecting all problems connected with sustainable development. The main goal of such an approach is to guarantee a higher level of life quality and to secure the future of the next generations.

Measurement as a stage of instrument designing: data and methods

Sustainable finance, by its definition, is a complex subdiscipline of finance concerning many other fields. This complexity is expressed in the number of factors which may exist in this area and could have a significant impact (directly or indirectly) on the sustainable finance or instruments offered on the market. Such a situation the reason for market monitoring made by many institutions issuing, offering and trading these instruments. Taking into account the global scale of the problem of sustainable finance is one of the priorities of the European Commission. Such guidelines and policies established at the global level set market development directions and financial products offered. Other issues which play an important role when designing sustainable offerings are aspects enabling the analysis and evaluation of events influencing sustainable development. It is closely tied with measurement and statistical data concerning this field. Two kinds of data are available: quantitative and qualitative. The problem of measurement supports the process of financial institutions monitoring and reacting to market demand and its changes. This allows one to verify the market and its institutions function in regard to existing guidelines and policies. All these activities could not be possible without proper data and appropriate methods of measurement. Using appropriate methods means that they are adjusted to functions, which should be fulfilled by quantitative methods:

- Analytical-descriptive function – analyzing financial market processes with particular focus on factors connected to sustainable finance. Such analysis allows for identification, description and better knowledge of the market regarding sustainable development. For example, market analysis could be conducted on sustainable financial products, analysis of sustainable projects and initiatives consistent with sustainable capital market policy or public companies' analysis from the perspectives of implementation of corporate social responsibility (CSR) policy. This function requires historical data.
- Diagnostic-control function – also called as analytical function with norm. Besides providing analysis similar to the first function, the diagnosis of the state is conducted here. One of the possible group of norms could be, for example, European Union guidelines.
- Prognostic function – it focuses on predicting the future levels of analyzed factors or future directions and trends of development of the analyzed variable. For example, this function may include forecasting trends and changes of needs in the functioning of sustainable finances.

The implementation of the functions of quantitative methods is important not only from measuring issues related to sustainable finances. The functions of quantitative methods also occupy an important place in the process of creating financial instruments (products) or business models for sustainable finance. This area of adaptation of the functions of quantitative methods (or quantitative methods in general) can be included in two steps or stages:

1   market analysis and assessment (if necessary, together with the forecast) from the perspective of sustainable finance factors. Here you can see, e.g. answers to the questions: How is it? What is the state of the financial market? What is the market in this? What is the market risk? What are the financial instruments (products) securing the ESG sphere? How do these instruments work?

2   based on the first stage, it becomes possible to identify gaps in the functioning of existing financial products or business models in sustainable finance. This may concern, for example, factors in the area of sustainable finance that were not previously included in the design of a given financial tool or which should be considered from the perspective of the functioning of sustainable finance. The identification may also refer to the existing gap in the demand for financial instruments and products, including ESG and concepts for the creation of new financial products.

Both the first and second stages should consider global guidelines on sustainable finance. An example here may be the fact that in the company's assessment from the perspective of fundamental analysis methods – ratio analysis – there are still no financial indicators that take into account, for example, ESG-related expenses. In this respect, there is also a gap from the perspective of global analysis, assessment and comparison of entities (e.g. companies from different markets). The same situation applies to the existing indicator of fundamental strength, which allows one to assess the fundamental strength of economic entities. In its construction, while the methodology of the indicator mentions the importance of the CSR area, it is difficult to include the financial indicator that would refer to it. It should be noted that the functions of quantitative methods are in the background of the measurement methods used or tools that can be used to build new or modify existing financial products for sustainable finance. However, they are important.

In practice, the measurement of effectiveness appears on every stage of its development. A three-stage conception indicates the following stages [Crawford & Di Benedetto, 2000]:

- the arrangement of the concept of a new product (demand identification, concept creation, preliminary identification),
- the transition of a concept into a product (final identification, prototype creation, an arrangement of technologies and production, market researches),
- start-up and commercialization (including production).

The first phase includes all activities leading to an arrangement of the conception to its technical development (it includes studies on the demand for the product, taking into account preliminary assessment market, technical and financial). The second phase is focused on the technical design of the product and on technologies. The third phase leans on the product/service introduction to the market. Almost all these phases concern designing and offering sustainable financial instruments to the market.

The number of factors appearing in sustainable finance require systematization. Generally, they could be divided into two types [Tarczynska-Luniewska & Tarczynski, 2015]:

- quantitative factors – quantifiable, allowing to make analysis, diagnosis or forecasting of the research problem using numerical values. Some of these factors are interrelated and remain in these relations, for example, expenditure on realizing the sustainable policy of the company and its rating level or levels of rates of return and volume of trading of RESPECT index. Commonly, quantitative variables are treated objectively.
- qualitative factors – factors which couldn't be expressed in numerical form. Such kind of data have a qualitative description and a subjective character. For example, taking into account sustainable finance, the data set could contain such kinds of variables as conducting business relations with countries violating human rights or cooperating with and supporting organizations working in the field of wilderness preservation. Limiting to the capital market frame (e.g. to the Warsaw Stock Exchange), qualitative data could be promoting the idea of corporate governance by rules of best practice for stock exchange companies.

It should be noted that both groups of factors are equally important. Qualitative factors are often indirectly reflected in the quantitative factors and vice versa (e.g. variable expenditures on the education and promotion of environmental protection represent the engagement of the company in terms of education and promoting a clean economy). Good practices of companies and consistency with ESG policy may be the basis for the adopted investment strategy in the case of sustainable investments, taking into account both creating and using financial instruments. A very important element of designing a financial instrument process is risk analysis. Detailed analysis of risk is possible thanks to the information provided by quantitative and qualitative factors. This also applies to the risks that may affect the shape of the implementation of sustainable development policy and sustainable finances (e.g. the risk of climate change, risk related to the functioning of weather financial instruments). Regardless of the approach (quantitative or qualitative) to the factors that may be faced by different market stakeholder groups, it is important to use appropriate measurement methods that allow analyzing and assessing the problem in all its areas. Regarding economic analysis, every factor describing economic activity should be treated as statistical data. Data should characterize some features, which allow one to properly use the appropriate methods. There are following features:

- availability – data should be available; it is possibly paid and free access to data (e.g. EU, Statistical Office, reports of listed companies),

- reliability – data come from reliable data sources (e.g. EU, Statistical Office, reports of listed companies),
- method of presentation – data require presentation in a relevant way, according to the requirements of the methods used, e.g. cross-sectional studies require presentation in cross-sectional series.
- comparability – data should be comparable or should have a possibility to be reduced to comparability (applied methods of reducing to comparability could not result in the loss of information).

One of the most important issues connected to measurement is the statistical quality of the data. Inseparably associated with quality is the problem of measurement scales, which influence the presentation of data and may determine the possibility of using analytical methods. The problem of data is important both from the existing measurement methods and tools, tools modification, as well as the creation of new methods and tools to make measurement possible. All these elements become more important when a new financial instrument or a new financial indicator is constructed within the needs generated by the market. Since the problem of sustainable finance became a global problem, it became more important to focus attention on the universality and utilitarian character of such a tool and its consistency with general directives, e.g. the capital markets union. Here arises a gap in the scope of measurement tools and financial instruments for sustainable finance, particularly in financial markets. Commonly used financial indicators often used in business models or in assessing the fundamental strength of companies to invest do not include ESG elements in their methodology. There is a lack of measures considering non-financial factors, which are an important element in assessing listed companies respecting sustainable finance policies. This is a very difficult issue requiring development and adoption of consistent guidelines for the comparability of regions, markets or entities. Financial aspects should include building, promoting and using indicators derived from international financial reporting standards for the assessment of measurable effects and, e.g. from Principles of Responsible Investing for non-measurable effects.

Conclusions

Designing instruments for sustainable finance and investment is a complex issue. First of all, this requires accepting the needs of doing business while respecting ethical, social and environmental goals. This applies to business entities at various levels of aggregation. From a longer perspective, such actions will also consistently apply to the market as a whole, including the financial market. The second issue is the social acceptance of goals by all groups of stakeholders, including employees, and their identification with the goals of sustainable development. Only then is it possible to apply sustainable mechanisms to business or more broadly to the market. Sometimes companies are only interested in one dimension of sustainable policy (ethical, social, cultural or environmental). This does not mean that there is no sustainable business – the company develops sustainably, but not in all areas. The ideal situation is to maintain all dimensions, but at every step improving the quality of life of society should be appreciated. This situation highlights the fact that there is a gap in financial products (instruments) for sustainable finance. In the process of designing financial instruments, therefore, there must be greater awareness when considering ESG factors. Existing financial instruments (or business models) on the market that do not hedge these issues should be modified in this respect. New products, in turn, are to consider ESG factors from the very beginning of their creation. There may be a problem in this area related to the unification of the approach used or rational factors related to ESG. Therefore, when building financial products, top-down guidelines should be taken into account, e.g. from the European Union and the UN, including the expert group established by the European Commission. The creation of financial instruments should be methodically and methodologically consistent. A lack of coherence and unification may prove harmful and pose threats to sustainable finances. This can include the following:

- the negative impact of badly constructed and used financial products or business models on the stability of the financial market or the wider economy (with local and global scale),
- incorrect assessment of the market, which may result in the wrong direction of capital allocation and flow (imbalance in the allocation of funds),
- incorrect targeting of investments in ESG-supporting technologies.

It should be remembered that the lack of financial instruments or the use of badly constructed instruments will increase the risk of sustainable finance functioning or organizations using this type of product.

An analysis and assessment of financial instruments or business models in retrospect will allow assessing the extent to which the phenomenon of sustainability and sustainable development in the field of finance functions. To make this possible, however, every effort should be made for a multifaceted approach to the construction of this type of tool.

References

Abraham, I. (2010). *Riba and recognition: Religion, finance and multiculturalism.* Australian Association for the Study of Religions Annual Conference Proceedings, Sydney, January.

Bocken, N.M., Short, S.W., Rana, P., & Evans, S. (2014). A literature and practice review to develop sustainable business model archetypes. *Journal of Cleaner Production*, 65, 42–56, https://doi.org/10.1016/j.jclepro.2013.11.039.

Crawford, C.M., & Di Benedetto, C.A. (2000). *New products management*. 6th ed. Irwin McGraw Hill, Boston.

Dasgupta, P. (2007). The idea of sustainable development. *Sustainability Science*, 2(1), 5–11, https://doi.org/10.1007/s11625-007-0024-y.

Dikau, S., & Volz, U. (2018). Central banking, climate change and green finance, https://eprints.soas.ac.uk/id/eprint/26445.

Eyraud, L., Clements, B., & Wane, A. (2013). Green investment: Trends and determinants. *Energy Policy*, 60, 852–865, https://doi.org/10.1016/j.enpol.2013.04.039.

Freeman, R.E. (1984). *Strategic management: A stakeholder approach*. Pitman, Google Scholar, Boston.

Fulton, M., Kahn, B., & Sharples, C. (2012). Sustainable investing: Establishing long-term value and performance. SSRN 2222740.

Krosinsky, C., Robins, N., & Viederman, S. (2011). *Evolutions in sustainable investing: strategies, funds and thought leadership* (Vol. 618). John Wiley & Sons, New Jersey.

Louche, C., Arenas, D., & Van Cranenburgh, K.C. (2012). From preaching to investing: Attitudes of religious organisations towards responsible investment. *Journal of Business Ethics*, 110(3), 301–320, https://doi.org/10.1007/s10551-011-1155-8.

Majoch, A.A., Hoepner, A.G., & Hebb, T. (2017). Sources of stakeholder salience in the responsible investment movement: Why do investors sign the principles for responsible investment? *Journal of Business Ethics*, 140(4), 723–741, https://doi.org/10.1007/s10551-016-3057-2.

Mansley, M. (2000). *Socially responsible investment: A guide for pension funds and institutional investors*. Monitor Press, Sudbury, Canada.

Mills, L., & Byrne, J. (2016). *Global trends in clean energy investment*. Bloomberg, New Energy Finance, http://www.bbhub.io/bnef/sites/4/2016/04/BNEF-Clean-energy-investment-Q1-2016-factpack.pdf.

Osterwalder, A., Pigneur, Y., & Tucci, C.L. (2005). Classifying business models: Origins, present, and future of the concept. *Communications of the Association for Information Systems*, 16, 1–25, https://doi.org/10.17705/1CAIS.01601.

Overview of European Sustainable Finance Labels. (2019). *Novethic*, June, https://www.novethic.com/sustainable-finance-trends/detail/.

Rasmussen, B. (2007). *Business models and the theory of the firm*. Working Paper No. 32. Victoria University of Technology, Melbourne, Australia, http://vuir.vu.edu.au/15947/1/15947.pdf.

Renneboog, L., & Spaenjers, C. (2012). Religion, economic attitudes, and household finance. *Oxford Economic Papers*, 64(1), 103–127, https://doi.org/10.1093/oep/gpr025.

Sandberg, J., Juravle, C., Hedesstrom, T.M., & Hamilton, I. (2009). The heterogeneity of socially responsible investment. *Journal of Business Ethics*, 87, 519, https://doi.org/10.1007/s10551-008-9956-0.

Stern, N., & Stern, N.H. (2007). *The economics of climate change: The Stern review*. Cambridge University Press, Cambridge.

Tarczynska-Luniewska, M., & Tarczynski, W. (2015). The method for detecting the leading sector on WSE. *Transformations in Business & Economics*, 14, 470–483.

United Nations UNCTAD. (2019). *Leveraging the potential of ESG ETFs for sustainable development*. Working Paper, November.

US SIF Foundation. (2018). *Report on US sustainable*. Responsible and Impact Investing Trends 2018.

Wilson, R. (1997). *Economics, ethics and religion: Jewish, Christian and Muslim economic thought*. MacMillan Press Ltd, London.

Chapter 6

Sustainable insurance

Agnieszka Majewska

Introduction

The insurance industry is one of the key pillars of the financial system. It is an important component of the economy by the value of the premiums it collects and the scale of its investment. Furthermore, it plays an essential social and economic role for citizens by covering their personal and business risks. It is possible to reduce their costs and assist them in case of an emergency by sharing risks over many participants. Providing reliable and quality products and services should be the aim of the business of insurance companies. This requires cooperation both with clients and business partners. It is not possible without understanding risk management.

The insurers have to consider a wide range of possible risks because the business is based on dealing with uncertainty. There are many sources of risk. Some of them are related to all firms, such as market risk, interest rate risk, operational risk, liquidity risk, environmental risk, reputation risk, credit risk, country risk, and law risk, and others are only linked to the insurance industry, e.g. catastrophe risk, mortality risk, and longevity risk. The risk profile of insurers has become increasingly complex since the new millennium. They have started offering riskier products such as shadow insurance, securities lending, and derivatives [Koijen and Yogo, 2017] and investing in riskier assets [Becker and Ivashina, 2015]. There is also some evidence from the last financial crisis which suggests that undertaking financial difficulties was not only due to using complex financial products but also due to the significant risk mismatch, especially in the life insurance industry [Peirce, 2014, McDonald and Paulson, 2015]. A key insight from the last financial crisis is that using sophisticated capital management tools needs more transparency and comprehensibility for all users. Therefore, one of the activities undertaken by insurers is implementing the principles of sustainability insurance.

This chapter is focused on the role of the insurance industry in sustainable development and provides special knowledge about sustainable insurance. All actions should be conducted under the general guidelines of the

United Nations Environment Programme Finance Initiative (UNEP FI), i.e. the Principles for Sustainable Insurance (PSI). Therefore, the PSI are presented at the beginning of the chapter. These principles are tailored to the needs and requirements of the insurance industry, the clients, and citizens. Then reference is made to environmental, social, and governance (ESG) risk and a sustainable insurance market. Areas covering actions to manage ESG risks in the insurance business are shown. The third section of the chapter refers to the crucial particulars of supervisors and regulators in insurance for selected countries. At the end, sustainable insurance empirical evidence and case studies are provided. Leading insurers who take into account sustainable development in their activities are also discussed.

The Principles for Sustainable Insurance

Research studies conducted by the UNEP FI from 2006 to 2009 started the initiative to develop the PSI. The research included consultations with representatives from the insurance industry, government and regulators, intergovernmental and non-governmental organisations, business and industry associations, academia and the scientific community. They covered Europe, Africa, Asia, Latin America and the Caribbean, Middle East, North America, and Oceania. The investigation focused on risks and opportunities in the insurance business associated with ESG issues [UNEP Finance Initiative, 2007].

The final PSI were established at the Rio+20 Summit in June 2012 by UNEP FI [UNEP Finance Initiative, 2012]. Sustainable insurance is defined as

> a strategic approach where all activities in the insurance value chain, including interactions with stakeholders, are done in a responsible and forward-looking way by identifying, assessing, managing and monitoring risks and opportunities associated with environmental, social and governance issues.
>
> [Scordis et al., 2014]

It is closely related to sustainable finance, which supports investors, companies, and issuers to take into account environmental and social considerations in making investment decisions. Sustainable finance is one of the tools which is used to pursue principles for sustainability insurance.

The PSI are a framework of developing and expanding innovative insurance solutions and risk management by insurance companies. The aims of the sustainable insurance formulated in this document are:

• reducing risk,
• developing innovative solutions,

- improving business performance,
- contributing to environmental, social, and economic sustainability.

The insurers should consider those objectives in their activities.

Although many of them have increased their focus on sustainable development, they are chiefly concentrated on financial risk management [Mills, 2007]. It is recommended to use the holistic approach which includes asset management in addition to financial and physical risk management. The triple role of the insurance industry in sustainable development is presented in Figure 6.1.

The following guidelines have been provided to achieve these purposes:

Principle 1. Embedding in decision-making ESG issues relevant to an insurance business.

Principle 2. Working together with clients and business partners to raise awareness of ESG issues, manage risk, and develop solutions.

Principle 3. Working together with governments, regulators, and other key stakeholders to promote widespread action across society on ESG issues.

Principle 4. Demonstrate accountability and transparency in regularly disclosing publicly the progress in implementing the principles.

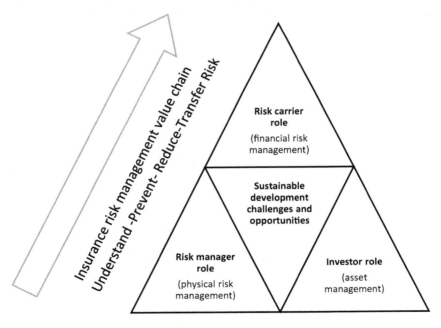

Figure 6.1 The triple role of the insurance industry in sustainable development
Source: UNEP FI [Bacani B., 2017].

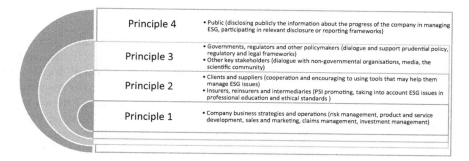

Figure 6.2 prefix:

Principle 4	• Public (disclosing publicly the information about the progress of the company in managing ESG, participating in relevant disclosure or reporting frameworks)
Principle 3	• Governments, regulators and other policymakers (dialogue and support prudential policy, regulatory and legal frameworks) • Other key stakeholders (dialogue with non-governmental organisations, media, the scientific community)
Principle 2	• Clients and suppliers (cooperation and encouraging to using tools that may help them manage ESG issues) • Insurers, reinsurers and intermediaries (PSI promoting, taking into account ESG issues in professional education and ethical standards)
Principle 1	• Company business strategies and operations (risk management, product and service development, sales and marketing, claims management, investment management)

Figure 6.2 The Principles for Sustainable Insurance: the areas of possible action
Source: Author's study based on [UNEP Finance Initiative, 2012].

The areas of possible actions under the Principles for Sustainable Insurance are given in Figure 6.2.

The principles are voluntary, so they are not designed to impose legal or regulatory sanctions or any form of claim towards any signatory's stakeholders or any third party.

Today one-quarter of the world's insurers are actively pursuing the enacted PSI. They join this global initiative to promote economic, environmental, and social sustainability. In November 2018 UNEP FI announced a partnership with the world's largest insurers which represent around 10% of world premium and USD 5 trillion in assets under management. These insurers include the following companies: Allianz (Germany), AXA (France), IAG (Australia), Intact Financial Corporation (Canada), Länsförsäkringar Sak (Sweden), MAPFRE (Spain), MS&AD (Japan), Munich Re (Germany), NN Group (Netherlands), QBE (Australia), Sompo Japan Nipponkoa (Japan), Storebrand (Norway), Swiss Re (Switzerland), TD Insurance (Canada), The Co-operators (Canada), and Tokio Marine & Nichido (Japan). The group works with leading banks and investors in order to advance financial sector know-how on climate change and develop a new generation of risk assessment tools.

ESG risk versus the sustainable insurance market

The PSI provide a global framework for the insurance industry to manage ESG issues. The insurance industry plays an important role in enabling a healthy, safe, resilient, and sustainable society. One of its purposes is to better understand, prevent, and reduce ESG risks. ESG issues also present a risk to insurance. Therefore, the UNEP Finance Initiative in cooperation with chief executive officers (CEOs) of leading insurance companies has prepared the ESG guide for the global insurance industry [UN Environment Programme's Principles for Sustainable Insurance Initiative, 2020].

There is no single approach to ESG risk management because it may differ for a country, a type of coverage, a line of business, clients, and other factors. Each insurance company undertakes its own decision regarding which ESG risk needs to be focused on. ESG issues represent a challenge for them to develop their internal processes and create a sustainable insurance market. Different motivations of implementing ESG issues by insurance companies exist. They want to keep their reputation and have an impact on employee morale and investor perception. On the other hand, each organisation has to consider its financial and strategic objectives to create an appropriate exposure of ESG risks. There are also societal norms, traditions, and the culture which determine the ESG risk aversion.

The UN Environment Programme's Principles for Sustainable Insurance Initiative [2020] points out eight areas covering possible actions to manage ESG risks in the insurance business. They are presented in Table 6.1.

It should be noted that ESG issues have an increasing impact on human life and the physical world. The influence of environmental factors on property and casualty (P&C) is obvious. Moreover, they also affect the health of the population in the long run. This means that the cost of health and sickness benefits are rising. This makes the value and cost of annuity policies also increase. Therefore, ESG factors are very important in the assessment of the risks to insurers' assets and liabilities. Nowadays, due to COVID-19, the healthy risk is especially important. Lockdowns and shutdowns affect both the economy and society. Besides dangers to public health, the pandemic could have long-lasting effects on people. Inequality, mental health problems, and lack of societal cohesion are intensifying [World Economic Forum, 2020]. Furthermore, the need for treatment and mortality trends may contribute to increasing health insurance claims. The worldwide pandemic has shown the weaknesses of health systems and pointed out the unreadiness in terms of pandemic preparedness across the world. It has highlighted the necessity for more fundamental investment in health and revealed the need for greater investment in other socio-economic priorities [World Economic Forum, 2020].

Supervisors and regulators in sustainable insurance

Supervisory and regulatory institutions have made some changes in approaching sustainability insurance issues over the last few years. Their

Table 6.1 Areas to manage ESG risks by insurance companies

No.	Area	Possible action
1.	Developing the ESG approach	Recognising three types of risk: a potential ESG risk, a potential elevated risk, and a potential high or direct risk. This classification is an indication and is based on the accumulated input from both the global consultation process and from team members of the organisation.
2.	Establishing the ESG risk appetite	Understanding the ESG risk aversion in decision-making. Indicate reputation and ethical perspectives by a company in deciding which ESG risks it wishes to focus on.
3.	Integrating ESG issues into the organization	Indicates a governance framework and internal guidance which enable ESG issues to be effectively integrated. Establishing the risk detection and prioritisation process, guidance on managing the risks, and the process for decision-making.
4.	Establishing roles and responsibilities for ESG issues	Indication of who is responsible for ESG issues. The support of the CEO, senior executives, and board members is necessary. Inclusion of other persons who can play an important role in the ESG decision-making process (e.g. insurance associations).
5.	Escalating ESG risks to decision-makers	Defining the escalation route to the ESG decision-making. The route of escalation must be clear from local levels up to senior-level management to avoid overburdening decision-makers.
6.	Detecting and analysing ESG risks	Implementation of tools to support the process of ESG risk management. Estimation of the budget to procure specialist tools or research which can support the detection of ESG risks.
7.	Decision-making on ESG risks	Examining the severity and frequency of ESG risks in the organisation. Indication of acceptable requirements to ESG risk mitigation. Implementation of an engagement process when an ESG risk is detected with clients or intermediaries. This activity supports clients and intermediaries in proactively managing their ESG risks.
8.	Reporting on ESG risks	Providing clear and public communication on ESG risks. Implementation of external auditing to verify the ESG risk management. It is evidence of a robust ESG risk management system for stakeholders.

Source: Author's study based on [UN Environment Programme's Principles for Sustainable Insurance Initiative, 2020].

activities mainly cover enhancing corporate and financial-sector disclosure of climate risks, assessment of systemic risks, susceptibility, and resilience to natural hazards, extending access to insurance [Sustainable Insurance, 2017]. Therefore, in 2016 the Sustainable Insurance Forum (SIF) was launched to construct the platform to provide engagement with leading regulators and to undertake a global consultation in insurance regulation, supervision, and policy. The following presents examples of activities taken by supervisors and regulators in selected countries.

United States

The world's first climate risk disclosure survey for insurance was made in the United States in 2009. It was conducted by the National Association of Insurance Commissioners (NAIC). It is a standard-setting and regulatory support organisation created and governed by the chief insurance regulators from the 50 states. The overall purpose of the NAIC is to assist state regulators in the performance of their regulatory oversight function concerning the insurance industry.

One of the first initiatives of the NAIC was the revision of the Financial Condition Examiners Handbook in 2013. The climate change factors into national-level supervisory standards were implemented. The companies have to consider the impact of climate change risks in determining their investment strategy and monitoring the risks in their investment portfolio [McHale and Spivey, 2016]. Certain states have taken independent steps to implement a new disclosure regime for insurance companies.

The California Department of Insurance (CDI) under its Climate Risk Carbon Initiative (CRCI) has introduced the following requirements:

- reporting by California-licensed insurance companies on the amount of their thermal coal enterprise holdings and that they voluntarily divest from thermal coal enterprises,
- disclosing by insurers financial information of their investments in fossil fuel (thermal coal, oil, gas, and utilities).

The improvement of the Climate Risk Disclosure Survey is presented in Figure 6.3.

The aim of the NAIC Climate Risk Disclosure Survey is to provide regulators, insurers, investors, and other stakeholders with substantive information about the risk of insurers from climate change and the steps insurers are taking to respond to those challenges. The report of Ceres also offers recommendations for insurers and regulators to improve their management

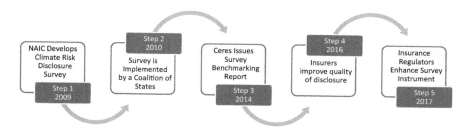

Figure 6.3 Development of insurer climate risk disclosure
Source: Messervy [2016].

and disclosure on wide-ranging climate risks [Messervy, 2016]. It should be underlined that public benchmarking of company responses made by Ceres in 2014 has contributed to a step-change in disclosure quality. Two general recommendations have been given to regulators:

1   Enhance the Climate Risk Disclosure Survey

   The survey instrument should be updated to better capture insurers' actual climate risk management performance. This means that regulators, investors, and other stakeholders have more detailed and timely information about what companies are doing, or are not doing, to address climate risk across their businesses.

2   Continue to Expand Climate Risk Disclosure

   In regard to the NAIC Climate Risk Disclosure Survey, in 2014 insurance regulators in six states (California, Connecticut, Minnesota, New Mexico, New York, and Washington) required the participation of insurers from 45 other domestic jurisdictions. They could advance the interests of their jurisdictions by requiring insurers under their purviews to provide information about climate risk management. Owing to this, regulators are able to assess companies' emerging risk strategies and outlooks for the future.

   The active actions on climate change challenges are also conducted by the Office of the Insurance Commissioner of Washington State (OICWS). The OICWS has formulated the guidance for other state regulators which should be used to evaluate insurers' climate change risks and investments during financial examinations.

Australia

In the insurance sector, the Australian Prudential Regulation Authority (APRA), which is the regulator of the financial services industry [APRA's policy priorities, 2018]:

• consulted on and finalised the application of APRA's risk management prudential standard to private health insurers,
• consulted on a framework to strengthen the role of the appointed actuary in insurers,
• initiated a review of aspects of the life reinsurance framework.

Nowadays, the work undertaken by APRA is used to modernise and change the existing requirements. APRA intends to update the following initiatives (see Table 6.2).

Brazil

In Brazil, the Superintendence of Private Insurance (SUSEP), who is directly linked to the Ministry of Finance, is the insurance regulator and commissioner. It is responsible for the supervision and control of the insurance and open private pension funds. The aim of its activity is also strengthening the public's confidence in the system and the development of a sound insurance market while complying with the international standards.

SUSEP is engaged in the modernisation of supervisory and regulatory procedures. It started to consult with regulated entities to obtain data and information about market practice on sustainability issues in November (the data were obtained from 75% of Brazil's 172 insurance companies). The investigation found that 80% of Brazilian insurance firms consider environmental issues to be important to their overall business strategy. However, few of them have implemented policies or mechanisms to consider and mitigate the impacts of climate change within underwriting policy, risk management, or investment decision-making [Sustainable Insurance, 2017]. Insurers are more focused on the product design side and on supplying insurance products which support low-carbon activities. Much lower activity is seen in the incorporation of ESG risks into an underwriting process and within the investment space.

In order to enhance the greater action on these priorities, SUSEP shall cooperate with the Brazilian insurance market association (the National Confederation of General Insurance, Private Pension and Life, Supplementary Health and Capitalization Companies – CNseg) and with the insurance firms. They are working to ensure that the issues addressed in the PSI are introduced in the insurance market to encourage risk management and innovation. They promote the integration of environmental risks into

Table 6.2 The key initiatives of APRA

The initiative	The objective	First half 2018	Second half 2018	2019
Prudential standard of risk management for the private health insurance (PHI) industry – CPS 220	Ensuring that all authorised deposit-taking institutions, general insurers, life companies, and private health insurers have in place appropriately structured and systematic approaches to risk identification and risk management	Implement		
Prudential standard of reinsurance management – LPS 230	Ensuring that reinsurance arrangements of life companies are subject to minimum standards of independent oversight by APRA	Finalise and implement	Finalise and implement	
The role of the appointed actuary	To improve the functioning of the appointed actuary role and to ensure that the role has the capacity to provide independent and unbiased advice and challenge in an efficient and effective manner	Finalise		Implement
Prudential standard LPS 117 – Capital Adequacy: Asset Concentration Risk Charge	Ensuring that life companies maintain adequate capital against the asset concentration risks associated with their activities	Consult	Consult	Finalise
Prudential standard LPS 114 – Capital Adequacy: Asset Risk Charge	To maintain adequate capital against the asset risks associated with insurance activities	Consult	Consult	Finalise

(Continued)

Table 6.2 (Continued)

The initiative	The objective	First half 2018	Second half 2018	2019
Prudential standard CPS 510 – Governance, Prudential standard CPS 520 – Fit and Proper, Prudential standard – HPS 310 Audit and Related Matters	Sound and prudent management by a competent board. The board makes reasonable and impartial business judgements in the best interests of the policyholders. Responsible risk management by the board. Establishment of requirements for the provision, to the board and senior management, of independent advice in relation to the operations, financial position, and risk controls of the business operations of the private health insurer.	Consult	Finalise	Implement

Source: Author's study based on APRA's papers [2016a, 2016b, 2017, 2018].

underwriting policy and incentivise green investments. Their activity is intended to affect the rising awareness of policyholders, intermediaries, and business partners about sustainability issues.

China

The enforcement of environmental regulations is made in China by including mandatory requirements for environmental pollution liability insurance. The programme started in 2013. The Ministry of Environmental Protection (MEP) and the China Insurance Regulatory Commission issued Guidelines on Pilot Projects of Compulsory Environmental Pollution Liability Insurance. It was directed to companies with highly polluting sectors at the beginning.

Revised in 2014, the Environment Protection Law points out the need for companies to purchase environmental pollution liability insurance to ensure compensation for environmental damages they cause.

Ongoing regulatory reforms and developments which are referred to sustainable insurance in China include [Bacani, 2015]:

- the China Risk-Oriented Solvency System,
- the construction of a national natural disaster risk-transfer programme and improvement of loss models and underlying data,
- the Scheme for the Overall Promotion of Life Microinsurance,
- microinsurance regulation in Taiwan, a province of China,
- the Treat Customers Fairly (TCF) Charter of the Hong Kong Monetary Authority,
- the establishment of the Independent Insurance Authority by the government of Hong Kong, Special Administrative Region, China.

United Kingdom

The Prudential Regulation Authority (PRA), part of the Bank of England, is responsible for the regulation of insurance firms in the UK. The Bank of England, through the PRA, realises prudential regulatory responsibilities and alongside its monetary policy and financial stability remits in the insurance industry. Two primary statutory objectives of the PRA are:

1   Promoting the safety and soundness of insurance firms,
2   Contributing to securing an appropriate degree of policyholder protection.

The secondary objective is promoting effective competition in the markets for services provided by PRA-authorised firms.

According to the 2008 UK Climate Change Act, PRA began to examine the impacts of climate change in the insurance industry in 2015. The investigation included around 500 insurers. The majority of them provide general insurance services, i.e. commercial, public liability, motor, and home insurance. In the smaller proportion are life insurance companies, and a small number of firms provide both general and life insurance products. In the report, which was the first document published by the PRA and Bank of England on the subject of a climate change, three types of risk related to the impact of a climate change on the insurance sector were indicated [Bank of England, 2015]:

1   Physical risk

   It refers to direct impacts from extreme weather events and natural disasters and to indirect impacts such as natural capital degradation or disruptions to trade. Increasing levels of physical risks may challenge general insurance liabilities. It is connected with the possibility of modelling of risk, which is often difficult to predict, especially in the long term.

2 Transition risk

This type of risk covers a range of potential developments, actions, and events related to the low carbon transition which could impact upon the safety and soundness of PRA-regulated insurance firms and their policyholders. It is especially associated with the financial risk stemming from changes in climate change–related policy and regulation, the rapid development of low-carbon technology, changing investor preferences, and significant developments in climate science which affect markets.

3 Liability risk

It refers to the liability of insurance which may increase from the costs of climate change damages being passed onto insurers through third-party liability policies. Suffering parties try to recover losses from others who they believe may have been responsible.

France

The French Energy Transition law, which was launched in August 2015, covers reporting requirements for financial institutions related to risks and opportunities with climate change. In conjunction with the provisions of Article 173, insurance companies are required to disclose information on their risk management of climate change consequences and to take into account their environmental footprint in their investment policy.

The criteria for applying implementation measures of these reporting requirements differ according to the size of entities. The entities with total assets below EUR500 million (including both the group and single entities) are obliged to disclose the following information:

- a full description of their investment policy related to ESG issues,
- a description of information practices of individual investors/subscribers on the ESG policy,
- a description of internal risk identification and management processes related to the ESG issues,
- information on a potential voluntary adherence to a specific code of conduct related to the ESG issues.

Larger entities with total assets above EUR500 million, in addition to the obligations mentioned earlier, are required to provide a detailed description of applied criteria. These include:

- the type of information computed: financial, non-financial information, internal or external analysis, etc.,
- the methodology which was applied, assumptions and results,

- the results of the analysis process and its impact on the investment policy and how it will contribute to the overall objective of limiting global warming.

The top award for the best investor climate-related disclosures contest was granted to AXA [AXA, 2016]. The AXA Group created a Responsible Investment Committee (RIC), which develops a global approach to responsible investing issues. It considers both reputation-related matters and the more positive inclusion of ESG issues in investment processes from both a performance and risk management perspective.

Referring to the regulations in the European Union, the most important is Solvency II. Starting in 2016, all insurers operating within the European Economic Area (EEA) are required to hold a capital buffer. This Solvency Capital Requirement (SCR) requires insurers to be sufficiently capitalised. It is intended to ensure they can withstand losses of a 1 in 200-year event, over a 1-year time horizon. Furthermore, insurers also have to consider a risk beyond this one-year time horizon as part of their Own Risk and Solvency Assessment (ORSA). It covers the potential impact of climate change as well. The European Solvency II regulatory regime is considered illustrative for other regulators. Many of the evolving regulatory capital regimes, for example, in the Asia-Pacific region, are conceptually similar to Solvency II and often apply comparable capital charges. Also, many US insurers and Bermudan reinsurers providing operations in Europe implement the provisions of Solvency II [BlackRock Global Insurance Report, 2019].

Sustainable insurance: empirical evidence and case studies

Increased interest in sustainable solutions in business taking also climate change into consideration is observed in various industries [Carrillo-Hermosilla et al., 2010; Machiba, 2010]. It implements climate factors into the existing business model and review of the ORSA to assess whether climate change factors are being appropriately identified and assessed [Sustainable Insurance, 2017]. Johannsdottir [2014] has proposed transforming the insurance linear business model to a closed-loop model for the purpose of integrating sustainability aspects into insurers' business models. The idea of transforming linear processes to closed-loop processes is taken from industrial ecology. The closed-loop model allows insurers to include environmental sustainability issues and gain from this.

Most leading insurers have integrated ESG considerations into their business models in the conviction that these may impact both risks and returns. They refer both to investment processes, ownership practices, and disclosing on climate risk management by insurers and the companies they insure and invest in. Following are presented business models of selected insurers who take into account sustainable development in their activities.

AXA

AXA, as an industry leader, feels an enormous responsibility for providing sustainable business activity. The corporate responsibility strategy of AXA, which covers ESG issues, is focused on the three following items:

1   Responding to climate change,
2   Health risk prevention,
3   Using responsible data.

The initiatives undertaken by AXA regarding climate change are presented in Table 6.3.

In the second area of sustainable strategy, AXA leads health risk prevention efforts. Through a series of strategic partnerships for health, AXA informs and educates the public about the risks related to health, traffic, and home risks. All its activities get people to make life safer for themselves and others by reducing the number and severity of accidents and injuries. Moreover, AXA does not make investments in sectors like tobacco.

The third area refers to using responsible data for the greater benefit of society. AXA analyzes risk-related data to better protect customers and strengthen IT security. Its initiative Give Data Back shares data that help

Table 6.3 The key initiatives of AXA's climate strategy

Area of action	Actions
Investments	– investing in green assets, divesting from certain carbon-intensive industries, – climate-related shareholder engagement – climate risk analysis with a long-term view
Insurance Underwriting	– underwriting restrictions on the coal and oil sands industries, "green/sustainable" products in both property and casualty and life and savings ranges – promotion of new insurance solutions designed for developing countries (typically parametric insurance)
Operations	– direct environmental footprint targets covering energy and carbon emissions, water, and paper
Thought leadership	– partnerships with non-governmental organisations – academic research (AXA Research Fund) – public/private partnerships to foster prevention
Biodiversity protection	– beginning to invest conserve, protect, and restore natural capital (land-based and marine) to preserve their ability to act as natural carbon sinks and high-value habitats necessary for conservation and biodiversity

Source: Author's study based on AXA Group [2019].

customers and society at large better understand risk and its management. The platform uses a transparent and collaborative approach which enables visitors to benefit from the company risk prevention experience. Unfortunately, the Give Data Back platform is no longer available in Europe. Nonetheless, there were some initiatives which have showed great success. These include, among others, exploring accident hotspots in Singapore, as well as sharing road data with public institutions in Mexico. Such activities encourage collaborative prevention.

It should be underlined that AXA, as a signatory of the UN PSI, works together with clients and business partners to raise awareness of ESG issues, manage risk, and develop solutions. The company has set up the AXA Atout Coeur (AXA Hearts in Action) programme. Employees of the company work on projects addressing key social and environmental issues around the world under this global programme.

Allianz

In order to provide sustainable insurance, Allianz Group provides activities in the following areas [Allianz Group, 2018]:

- as a sustainable insurer, it provides solutions that support sustainable development and a low-carbon future,
- as a responsible investor, it takes into account ESG issues in all investment decisions,
- as a trusted company, it promotes a culture of integrity and compliance with the law, protects the data and privacy of customers, and provides responsible sales by properly informing and advising customers,
- as an attractive employer, it engages all employees globally in delivering the renewal agenda,
- as a committed corporate citizen, it maximizes positive societal impact across all operating countries by the company and creates and encourages an enabling environment for social ventures.

The sustainable insurer and the responsible investor are two categories directly relating to providing business activity. Table 6.4 shows activities carried out by Allianz under those categories.

One of the most important activities mentioned earlier is developing products and services that create social value or support sustainable development. Those products and services make a positive environmental impact and consider climate-related concerns. In addition to actively contributing to environmental issue they impact societal improvements. Figure 6.4 shows the five main groups of sustainable insurance solutions.

The number of sustainable solutions implemented by Allianz grows steadily year by year. In 2018 it reached 185 and was higher by 16 compared

Table 6.4 Activities carried out by Allianz as a sustainable insurer and responsible investor

Category	Activities
Sustainable insurer	– Giving prevention and resilience advice for natural catastrophes to clients or societal partners – Offering social solutions – Ensuring fair treatment of claims – Integrating ESG criteria in insurance decisions – Offering green solutions – Providing insurance products in developing countries vulnerable to climate change – Extending digital offerings for customers
Responsible investor	– Integrating ESG criteria in investment decisions for funds managed for third parties – Offering socially responsible investing products for third-party clients – Investing in projects and companies with a positive social impact – Investing in forest protection – Investing in renewable energies – Integrating ESG criteria in investment decisions for insurance assets

Source: Author's study based on Allianz Group [2018].

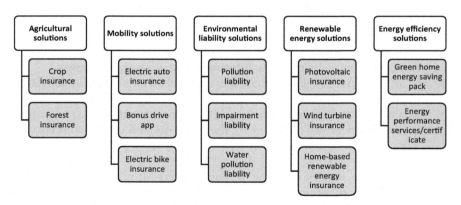

Figure 6.4 Products with an environmental or social added value

Source: Allianz Group [2018].

to the previous year. The most significant increase (nearly twofold) was recorded in implementing new sustainable asset management solutions. Sustainable insurance solutions have increased slightly, while the number of sustainability solutions in other areas has remained constant.

American International Group, Inc.

American International Group, Inc. (AIG) is the next of many insurers responsible for sustainable development. Like the previously presented companies, it systematically integrates ESG issues into its business activity. The company holds long-standing and leading positions on the US market. Therefore, in comparison to companies from Europe AIG carries out significant activities in the area of insurance against catastrophes and assimilated events. The company uses them as an opportunity to detect and remedy any weaknesses in how they model, select, and price risk. In 2018 AIG developed an internal US hurricane model. It was designed to provide a more bespoke view of the impacts of climate change on the wind, storm surge, and flood risks to business stemming from US hurricanes [AIG, 2019]. Furthermore, the company has used its internally developed flood models to create a view of flood maps in more than 25 countries. They are customised to incorporate climate change views for underwriting and pricing flood risk.

Besides internal models, AIG implements recognised third-party catastrophe risk models to evaluate and simulate the frequency and severity of catastrophic events and associated losses to its portfolio of exposures. It should be noted that these models are sophisticated and need to apply physically based modelling with parameters and assumptions derived from historical hazard and claims data. All models are continually improved and updated as new science and data become available.

The company analyses extreme disaster scenarios as well, for example, a flood in London as a result of the failure of the Thames Barrier due to an increase in windstorm intensity and sea-level rise from climate change. Extreme disaster scenarios enable monitoring and management of concentration risks within its portfolios. Furthermore, by using multiple levels of risk management processes, AIG is able to anticipate potential changes to its risk profile, pricing models, and strategic planning.

AIG cooperates actively with the scientific community and disseminates scientific information that climate change is a reality of increasing global concern. In 2006, the company was one of the first US insurers to formally recognise anthropogenic climate change, indicated by higher concentrations of greenhouse gases, a warming atmosphere and oceans, diminished snow and ice, and rising sea levels [AIG, 2019].

Environmental issues and objectives are also implemented by AIG in the confines of the AIG ESG Dividend Fund launched in December 2016. It is an open-end fund which seeks to provide total return by investing in stocks selected through a systematic, four-step process. This approach covers ESG ratings, profitability, valuation, and dividend yield to select a focused portfolio of 40 dividend-paying stocks. Like all responsible companies, AIG strives for transparent and clear communication with its stakeholders and adheres to high ethical standards in financial and non-financial disclosures.

The business activities presented earlier, including sustainable development, represent examples for the leading insurers. They have been chosen on the grounds that were characterised by high total assets and being systemically important insurers. It should be underlined that more insurers work for sustainable development and have implemented ESG issues.

Conclusion

This work has documented that the insurance sector which covers protecting a different kind of risk for households – the economic, financial, and corporate undertakings – is closely related to ESG issues. In view of climate change, a degradation of the environment, and a growing number of natural disasters, insurers play an important role in enhancing the safety of society, resource efficiency, and education in sustainable development.

The chapter indicates that one of the most important activities is the effective implementation of PSI by more and more insurance companies. It is not possible without developing and expanding innovative solutions and risk management. The study has shown that insurance companies are increasingly oriented towards offering innovative products and services, considering ESG issues. The strategies of leading insurance companies which are presented in this work confirm the systematic integration of ESG factors into insurance activity. Moreover, they promote and raise awareness of ESG issues, manage risk, and develop sustainable solutions.

The chapter indicates that the role of supervisors and regulators is crucial to implement sustainable development principles in the insurance industry. Although the insurance industry is highly regulated and regulations are one of the factors influencing underwriting and terms of risk severity, the implementation of worldwide rules is needed. Therefore, global regulations are a key issue in managing ESG risk effectively and efficiently in the insurance business. It should be noted that there are also national regulatory frameworks, but the development of sustainable insurance needs the global cooperation of governments, national and world organisations, and regulatory agencies. Sustainability factors have to be effectively integrated into the regulation and supervision of insurance companies.

The long-term vision of developing sustainable insurance is closely related to the responsibility of insurers and their willingness to implement the PSI. Looking forward, the main ESG issues which are related to sustainable insurance and need to be implemented in the business activity of insurers are:

- increasing vulnerability to natural disasters,
- natural resource degradation,
- water scarcity,
- environmental pollution,

- ageing populations,
- emerging health risks related to globalisation (possibility of a pandemic – COVID-19),
- trust and reputation issues,
- accountability and transparency of business activity.

References

AIG (2019), 2018 Climate-Related Financial Disclosures Report, TCFD.

Allianz Group (2018), Shaping Our Sustainable Future, Sustainability Report 2018.

APRA (2016a), Risk Management Prudential Standard for Private Health Insurers, Discussion Paper.

APRA (2016b), The Role of the Appointed Actuary and Actuarial Advice Within Insurers, Discussion Paper.

APRA (2017), Risk Management Prudential Standard for Private Health Insurers, Response to Submissions.

APRA's Policy Priorities (2018), Information Papers, APRA 31 January, www.apra. gov.au/sites/default/files/policy_agenda_2018.pdf.

AXA Group (2016), Award on Investor Climate-Related Disclosures, TCFD.

AXA Group (2019), Climate Report, TCFD.

Bacani, B. (2015), A Systemic View of the Insurance Industry and Sustainable Development: International Developments and Implications for China, Chapter 9 in Greening China's Financial System, International Institute for Sustainable Development.

Bacani, B. (2017), Principles for Sustainable Insurance, Insuring for Sustainable Development, UNEP FI North American Members Meeting, New York, September.

Bank of England (2015), The Impact of Climate Change on the UK Insurance Sector, A Climate Change Adaptation Report by the Prudential Regulation Authority.

Becker, B., and Ivashina, V. (2015), Reaching for Yield in the Bond Market, *Journal of Finance*, 70(5), 1863–1902, https://doi.org/10.1111/jofi.12199.

BlackRock Global Insurance Report (2019), Re-Engineering for Resilience, EIU.

Carrillo-Hermosilla, J., del Río, P., and Könnöläc, T. (2010), Diversity of Eco-Innovations: Reflections from Selected Case Studies, *Journal of Cleaner Production*, 18(10–11), 1073–1083, https://doi.org/10.1016/j.jclepro.2010.02.014.

Insuring for Sustainability (2007), United Nations Environment Programme Finance Initiative, PSI, May.

Johannsdottir, L. (2014), Transforming the Linear Insurance Business Model to a Closed-Loop Insurance Model: A Case Study of Nordic Non-Life Insurers, *Journal of Cleaner Production*, 83, 341–355, https://doi.org/10.1016/j.jclepro.2014.07.010.

Koijen, R. S. J., and Yogo, M. (2017), Risks of Life Insurers: Recent Trends and Transmission Mechanisms, in F. Hufeld, R. S. J. Koijen, and C. Thimann, eds., *The Economics, Regulation, and Systemic Risk of Insurance Markets*, Chapter 4 (Oxford University Press, Oxford).

Machiba, T. (2010), Eco-Innovation for Enabling Resource Efficiency and Green Growth: Development of an Analytical Framework and Preliminary Analysis of

Industry and Policy Practices, *International Economics and Economic Policy*, 7(2–3), 357–370, DOI: 10.1007/s10368-010-0171-y.

McDonald, R. L., and Paulson, A. (2015). AIG in Hindsight, *Journal of Economic Perspectives*, 29(2), 81–106, DOI: 10.1257/jep.29.2.81.

McHale, C., and Spivey, R. (2016), Assets or Liabilities, Fossil Fuel Investments of Leading U.S. Insurers, Ceres.

Messervy, M. (2016), Insurer Climate Risk Disclosure Survey Report & Scorecard: 2016 Findings & Recommendations, Ceres.

Mills, E. (2007), From Risk to Opportunity: 2007: Insurer Responses to Climate Change, Ceres, http://insurance.lbl.gov/opportunities.html.

Peirce, H. (2014), Securities Lending and the Untold Story in the Collapse of AIG, Unpublished Manuscript, George Mason University.

Principles for Sustainable Insurance (2012), United Nations Environment Programme Finance Initiative, PSI, June.

Scordis, N. A., Suzawa, Y., Zwick, A., and Ruckner, L. (2014), Principles for Sustainable Insurance: Risk Management and Value, *Risk Management and Insurance Review*, 17(2), 265–276, DOI: 10.1111/rmir.12024.

Sustainable Insurance, The Emerging Agenda for Supervisors and Regulators (2017), United Nations Environment Programme, August.

UN Environment Programme's Principles for Sustainable Insurance Initiative (2020), Managing Environmental, Social and Governance Risks in Non-Life Insurance Business, June.

World Economic Forum (2020), COVID-19 Risks Outlook. A Preliminary Mapping and Its Implications, May.

Chapter 7

A sustainable capital market

Wesley Mendes

Who has been concerned with sustainability?

Around the world, newspaper headlines have recently featured large companies, but unfortunately these frequently haven't been for positive reasons (Mendes-Da-Silva, 2019; Hayat, 2015). The subjects of these headlines range from corruption involving top executives to environmental aggression of catastrophic proportions, including the worsening quality of life of society. In the field of corporate governance, the Enron scandal in 2001 in the United States and Operation Car Wash in 2015, related to the Brazilian state oil company Petrobrás (Watts, 2017), are illustrations of how the top level of administration can compromise the standards of corporate governance.

In the field of environmental disasters there is no lack of examples such as Deepwater Horizon (Roberto, 2011) and Vale[1] (Pearson et al., 2019). The fact is that performance deficits in terms of sustainability can motivate the reactions of players in the capital market, from consumers to investors (see Table 7.1). In this manner, boycotts of company products and even decisions not to allocate financial resources in investment portfolios can explain the loss of market value of a company that ignores aspects related to sustainability, i.e. environment, social and governance aspects (ESG), which are recognized as a mark of the new era in communications between companies and society (Norton, 2020).

Guaranteeing growth and economic development are two of the main objectives of every country (see Table 7.2). It's common for economists and researchers in business to traditionally consider factors such as capital, work and technology as the main factors of economic growth (Seyfang & Longhurst, 2013). In addition, recent studies offer causal evidence that investors value ESG factors in the capital market (Hartzmark & Sussman, 2019). The subprime financial crisis of 2008 demonstrated that there are substantial economic effects when there is a lack of trust and/or resilience in financial systems. More recently, this sentiment has been highlighted by the COVID-19 pandemic (Carlsson-Szlezak et al., 2020).

According to Correia et al. (2020), pandemics have an effect on economies. These authors point out that the 1918 influenza pandemic reduced manufacturing production by roughly 18% at the time. In the particular case of the recent coronavirus pandemic, it seems to permanently emphasize the relevance of sustainability concerns to the capital market – at least that is what recent economic literature and the international financial media have highlighted (Krosinsky, 2013; Garcia, 2020; Mackintosh, 2020). In essence, as a consequence of the possibility that the preferences of institutional investors and consumers may change as a result of this pandemic, factors related to the sustainability of the business are considerably more relevant than before the pandemic.

The coronavirus pandemic can increase the focus of investors on ESG aspects of companies, implying the prioritization of investments in certain sectors, such as infrastructure. The pandemic is likely to emphasize external pressures from consumers, investors and policymakers on companies. As a result, managers (executives and directors) will need to work to more clearly demonstrate how resilient their cash flow is to crises. During the coronavirus pandemic, for instance, there were reactions from different sectors of society against opportunistic behavior – for example, when suppliers took advantage of the shortage of face masks and the need for people to increase the frequency of hand washing. As a result, instead of prices being defined solely by supply and demand, suppliers were pressured to maintain prices basically at previous levels under the risk of being exposed to a shameful judgment on the part of society.

Likewise, many companies have gone beyond what is required by law to try to protect workers. At the same time, big companies, like Amazon.com, faced pressure from unions and foreign governments for virtually not doing enough (Mackintosh, 2020). ESG aspects, as a proxy for the firm's sustainability, therefore, due to the pandemic, may start to more explicitly affect both the cash generation and the firm's cost of capital in the context of the capital market.

Recently, the view has been disseminated that citizens are worried about their health and the health of their families, and above all the potential impacts on their means of subsistence. It has been revealed, therefore, that it is essential for leaders to emit a strong response designed to alleviate and reduce the uncertainty caused by COVID-19, especially among those most vulnerable (Carlsson-Szlezak et al., 2020). In essence, the short-term view has given way to the premise that political and business leaders should offer responses that deal with relevant issues in a manner that prioritizes health and social and economic equilibrium. As a result, they hope to promote economic resilience. The capital market plays a central role within this context.

A financial system that functions well allows an economy to fully explore its growth potential because it guarantees that the best investment opportunities will receive the necessary financing. This chapter seeks to discuss the role performed by the capital market in promoting sustainable development. The capital market is a very flexible tool to meet growing economic, social and environmental needs. Figure 7.1 illustrates associations between the size of capital markets in economies and sustainability issues.

Note that the capitalization of the market as a percentage of a country's economy is mainly associated with the promotion of pollution due to a reduction of protected areas, as well as a less intensive use of renewable

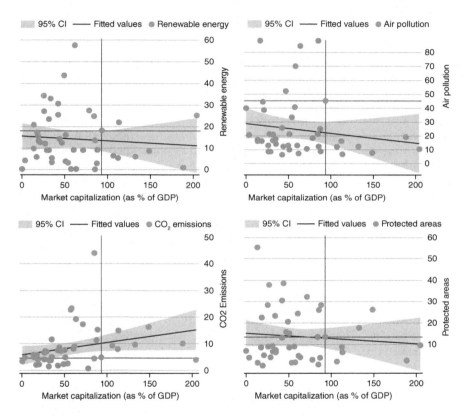

Figure 7.1 The relationship between market capitalization and country performance in sustainability metrics

Source: Prepared by the author based on data from the World Bank (2018).

Note: This graph presents the relationship between market capitalization (% of GDP) and four different sustainability metrics for various countries in 2018: renewable energy (% of total final energy consumption in 2015), air pollution (ambient PM2.5 air pollution, i.e. micrograms per cubic meter of mean annual exposure in 2016), protected areas (nationally protected terrestrial and marine areas, i.e. % of total area in 2016) and CO_2 emissions (carbon dioxide emissions, i.e. metric tons per capita in 2014). The solid vertical and horizontal lines mark, respectively, the references for world average market capitalization (in % of GDP) roughly 93%; the portion of energy consumption coming from renewable sources, air pollution, protected areas and CO_2 emissions (proceeding clockwise). The data tabulated for the construction of this graph are available at http://doi.org/10.5281/zenodo.3464558.

sources of energy. This scenario appears to point to a development agenda that is worthy of attention, bearing in mind the maintenance of health and social equilibrium in the face of capital market growth. In fact, the research community, as well as governments, international organizations, companies and investors (CalPERS, 2014; Eccles & Klimenko, 2019), appear to be more and more concerned with the production of studies which contemplate issues related to sustainability and the capital market. In regard to this, Table 7.1 presents ten studies with at least 100 citations as of March 2020, in the middle of the global COVID-19 catastrophe. Note that the works with the largest number of citations began to appear just before the turn of the century.

Bearing in mind that companies need resources (including financial resources) to develop their operational activities, it is assumed that individual and institutional investors have increasingly demanded attitudes in line with development oriented towards timeframes that extend beyond the shortest window (Green et al., 1996) and that sustainability issues may end up requiring new standards of industry behavior. This may include the utilization of clean energy and even the preservation of fauna and flora. For example, Eccles (2018) discusses the reasons why a New York–based hedge fund holding approximately US\$2 billion in Apple stock and the California State Teachers' Retirement System (CalSTRS) sent an open letter to the board of Apple outlining their concerns about the possibly undesirable effects of Apple products on child development.

The role of the capital market

One of the main reasons why companies look to the capital market is the chance to raise money at a lower cost, which may be subject to ESG aspects (Albarrak et al., 2019; Kim et al., 2015). This depends, however, on the capacity of these companies to remunerate investors, who prioritize companies who offer better return for risk. In this sense, the quality of institutional communication with the market, through reports containing precise, extensive and reliable information, is fundamental to companies (Mendes-Da-Silva & Onusic, 2014; Mendes-Da-Silva et al., 2014). Therefore, it's probable that the search for growing returns coexists with better ESG practices (Cheng et al., 2014; Orsato et al., 2015).

To the extent that companies manage to access the capital market, the benefits for society become viable, such as:

1   Individuals and families have investment alternatives to ensure their future power of consumption;
2   Consumer benefits: given that the maximization of share prices requires efficiency, companies seek greater quality at lower costs.

Table 7.1 Examples of the most cited articles in the sustainability and capital
market field

	Title	Authors	Year	Journal	# of citations
1	Capitals and capabilities: A framework for analyzing peasant viability, rural livelihoods and poverty	Bebbington	1999	World Development	983
2	Rethinking collaboration and partnership: A public policy perspective	Hall	1999	Journal of Sustainable Tourism	231
3	Playing with risk? Participant perceptions of risk and management implications in adventure tourism	Cater	2006	Tourism Management	172
4	Outreach, sustainability and leverage in monitored and peer-monitored lending	Conning	1999	Journal of Development Economics	165
5	Towards sustainable city policy: An economy-environment technology nexus	Camagni et al.	1998	Ecological Economics	149
6	Metals recycling: Economic and environmental implications	Ayres	1997	Resources, Conservation and Recycling	141
7	Community social responsibility and its consequences for family business performance	Niehm et al.	2008	Journal of Small Business Management	122
8	Social capital, development, and access to resources in highland Ecuador	Bebbington and Perreault	1999	Economic Geography	116
9	UK evidence on the market valuation of research and development expenditures	Green et al.	1996	Journal of Business Finance and Accounting	102

(Continued)

Table 7.1 (Continued)

	Title	Authors	Year	Journal	# of citations
10	Growing green money? Mapping community currencies for sustainable development	Seyfang and Longhurst	2013	Ecological Economics	100

Source: Prepared by the author based on data collected from the Scopus platform. Note: This table presents the 10 most cited articles out of 405 identified articles on March 27, 2020. The search used the terms 'sustainability and capital and markets' in the title, abstract or keywords of articles published in English in three fields of knowledge: i) business management and accounting and economics, ii) econometrics and finance and iii) decision sciences.

This implies that companies should develop products and services that customers want and need, inducing new technologies and products;

3   Employee benefits: in general, successful companies need more employees to achieve higher share prices.

In the search to attain benefits such as these, many governments are prioritizing the privatization of state-owned companies (Bachiller, 2017). It is expected, therefore, that privatized companies will have increased sales productivity, tending to grow and need new employees (Abramov et al., 2017).

What is sustainability within the context of the capital market?

The adjustment of the capital market to the sustainability paradigm is not restricted to conceptual strategies or approaches. Hartzmark and Sussman (2019) have found evidence through experiments that suggest that sustainability is a factor seen as a positive forecast of future performance. However, these authors did not find evidence that high-sustainability funds have better performance than low-sustainability funds. The evidence is consistent with positive effects that influence the expectations of sustainable performance for funds. While there are indications, according to Hong and Kacperczyk (2009), that companies that operate in sin industries, e.g. cigarettes, alcoholic beverages and gambling, perform better than other companies within the context of the capital market, there is recent news that technology companies dominate funds dedicated to ESG factors (Otani, 2020).

Frequently, funds that present themselves as sustainable investments are not necessarily focused on companies that fight climate change, develop wind turbines or prioritize diversity in their top levels of administration.

Instead of this, many of them basically prioritize a portfolio composed of big technology stocks. In the United States, companies focused on ESG aspects effectively associated with themes such as renewable energy, potable water and racial and gender diversity are relatively underrepresented in the portfolios of these funds.

Otani (2020) points out that, for example, while NextEra Energy, Inc. (www.nexteraenergy.com), a large operator of wind and solar parks, is not on the list of shares maintained by the Royal Bank of Canada (www.rbccm.com/en/) in ESG funds, technology companies such as Amazon.com and Facebook are. Therefore, it isn't uncommon to find that some of these funds invest in shares that do not conform to ESG aspects. As has been noticed by the international financial media, in the beginning of 2020, among the ten companies that received more capital investment in ESG funds, just three are not in technology. Thus, companies such as Microsoft are preferred due to their being classified as having superior performance in terms of data privacy and security, corporate governance, an absence of corporate corruption and instability, as well as their capacity for innovation using clean technologies. It may further be noted that Alphabet, Visa, Apple, Cisco, Mastercard and Adobe have drawn the attention of investors interested in companies which seek conformity with practices that respect ESG factors (Otani, 2020). However, being in technology is not enough by itself to receive elevated ESG ratings.

What rating agencies say about the relevance of ESG factors

During recent years, rating agencies of international prestige, such as Moody's, have announced their vision of the growing relevance of ESG aspects in the classification of credit risk within the context of the capital market. In addition, the view that it is necessary to examine the levels of ESG conformity in interaction with financial profile variables has been prevalent, e.g. a strong financial position and low financial leverage are important characteristics for managing environmental and social risks (Nauman & Gross, 2019).

In this way, there is an emerging vision that the credit rating industry tends to incorporate ESG aspects together with variables traditionally considered in the rating process. This trend has been reflected in the growing conception that corporate reports should contemplate ESG aspects to provide analysts with a more accurate vision of a company's expected performance (Flores et al., 2019). The Fitch Ratings Agency has launched an integrated point system to identify how ESG factors affect individual decisions in the classification of credit, according to the method adopted by this rating agency (Table 7.3). Fitch claims to be the only agency currently offering this level of granularity or transparency in terms of the impact of ESG factors on credit fundamentals.

Table 7.2 International Market Development Indicators (2018)

	Market Capitalization				Market liquidity^a		Turnover ratio^b		# domestic Co.^c		S&P Global^d	
	US$ Millions		% of GDP									
	2010	2018	2010	2018	2010	2018	2010	2018	2010	2018	2010	2018
Argentina	63,910	45,986	15.1	8.9	0.6	0.9	4	9.9	101	93	55.3	−41
Australia	1,454,491	1,262,800	126.9	88.2	98.7	54	77.8	61.3	1,913	2,004	12.5	−16
Austria	126,032	116,802	32.2	25.6	12.5	8.7	38.8	33.8	89	67	10.9	−26
Bahrain	20,060	21,863	78	57.9	1.1	2.6	1.4	4.4	44	43	10	−6.5
Bangladesh	41,617	77,391	36.1	28.2	4.2	5.9	11.6	..	192	593	37.6	−18
Belgium	268,726	321,094	55.6	60.4	22.2	..	40	..	161	111	0.5	−27
Brazil	1,545,566	916,824	70	49.1	41.1	41.2	58.8	83.9	373	334	6.5	−4.5
Canada	2,171,195	1,937,903	134.6	113.2	87.1	80.2	64.7	70.9	3,771	3,330	22	−19
Chile	341,799	250,740	156.4	84.1	26.5	14.6	16.9	17.4	227	205	47.2	−20
China	4,027,840	6,324,880	66.2	46.5	136	96.1	205	207	2,063	3,584	6.9	−21
Hong Kong SAR, China	2,711,316	3,819,215	1,185.90	1,052.10	651	625	54.9	59.4	1,396	2,161	21.3	−12
Colombia	208,502	103,848	72.9	31.4	7.9	4.1	10.9	13.2	84	66	44.1	−19
Croatia	25,596	20,509	42.8	33.7	1.8	0.4	4.1	1.2	240	127	−0.4	−9.3
Cyprus	6,834	3,313	26.7	13.5	2.9	0.2	10.9	1.7	110	91	−72	−20
Egypt, Arab Rep.	84,277	42,006	38.5	16.7	17	5.8	44.2	34.5	227	250	11.5	−10
France	1,911,515	2,365,950	72.3	85.2	51.1	..	70.6	..	617	457	−9.9	−11
Germany	1,429,719	1,755,173	41.8	43.9	43.7	40.4	104.5	92.1	690	465	7.4	−18
Greece	67,586	38,371	22.6	17.6	13.6	5.5	60.1	31	277	183	−44	−32
Hungary	27,708	28,935	21.2	18.6	20.2	6.4	95.5	34.2	48	43	−11	−8.5
India	1,631,830	2,083,483	97.4	76.4	64.5	46.2	66.2	58.1	5,034	5,065	18.7	−11
Indonesia	360,388	486,766	47.7	46.7	13.8	10	29	21.5	420	619	37.9	−9.5
Ireland	60,368	110,154	27.2	28.8	4	8.4	14.7	29.1	50	43	−7.7	−22
Israel	227,614	187,466	97.4	50.7	46.3	16.9	47.6	33.3	596	420	7.4	−3.8
Japan	3,827,774	5,296,811	67.2	106.6	74.9	127	111.6	119	2,281	3,652	9.6	−12

Jordan	30,864	22,740	116.2	53.8	32.4	5.6	27.9	10.4	277	195	-8.6	-0.4
Kazakhstan	26,673	37,005	18	21.7	0.4	0.7	2	3.4	61	97	-1	-38
Korea, Rep.	1,091,911	1,413,717	99.8	87.3	149	152	149.3	174	1,781	2,186	25.3	-22
Lebanon	12,697	9,675	33	17.1	4.9	0.7	14.8	..	10	10	-8.7	-14
Luxembourg	101,129	49,483	190	71.2	0.3	0.1	0.2	0.2	29	27	-3.3	-22
Malaysia	408,689	398,019	160.3	112.3	45	38.2	28.1	34	948	902	35.1	-13
Mauritius	7,753	9,848	77.5	69.2	3.6	3.2	4.7	4.6	62	99	8.2	-4.2
Mexico	454,345	385,051	43	31.5	10.5	7.7	24.5	24.4	130	140	26.6	-16
Morocco	69,152	61,081	74.2	51.5	6.5	3.3	8.8	6.4	73	75	13.1	-11
Namibia	1,176	2,462	10.4	17	0.2	..	1,721.50	..	7	10	24.2	-12
Netherlands	661,099	1,100,105	78.1	132.3	65.8	..	84.3	..	150	103	1.2	-15
New Zealand	43,516	86,133	29.7	42	1.3	5.9	4.4	14.1	137	131	5.2	-4.1
Nigeria	50,546	31,521	13.9	7.9	1.4	0.7	10.1	8.2	215	164	20.3	-18
Norway	295,288	267,382	68.8	61.5	49.4	27.4	71.8	44.6	195	186	13.7	-11
Oman	28,316	18,782	48.3	23.7	5.6	2.4	11.7	10.2	114	110	12.2	-10
Pakistan	38,007	91,864	21.4	33	6.6	9.9	30.7	30	644	:	15.3	-33
Panama	8,348	15,648	28.4	24.1	0.2	..	0.8	..	34	30	12.8	-16
Peru	103,347	93,385	70.1	42	2.7	1	3.9	2.4	199	211	51.3	-1.7
Philippines	157,321	258,156	78.8	78	11.1	8.8	14.1	11.3	251	264	56.7	-17
Poland	190,706	160,483	39.8	27.4	14.5	9.3	36.4	34.1	570	823	11.3	-17
Portugal	81,997	61,934	34.4	26	22.5	..	65.4	..	52	40	-17	-15
Qatar	76,531	163,047	66.4	84.9	14.7	9.9	63	11.6	43	46	27.7	22.2
Russian Federation	951,296	576,116	62.4	34.8	33.2	8.9	53.3	25.5	556	221	21.7	-9.4
Saudi Arabia	353,410	496,353	66.9	63.4	38.1	29.3	56.9	46.3	146	200	9	10.7
Singapore	647,226	687,257	269.9	188.7	128	60.3	47.2	31.9	461	482	18.4	-14
Slovenia	9,428	7,267	19.6	13.4	1	0.7	5.1	6	72	31	-20	-7.2
South Africa	925,007	865,328	246.4	235	73.9	80.1	30	34.1	352	289	32.1	-26
Spain	1,171,625	723,691	81.8	50.7	96	43.7	117.3	86.2	3,310	2,979	-25	-19
Sri Lanka	19,924	15,575	35.1	17.5	8.8	1.2	25.1	7	241	297	84.6	-24
Switzerland	1,229,357	1,441,160	210.6	204.3	150	133	71.2	65.1	246	236	11	-12
Thailand	277,732	500,741	81.4	99.2	65.2	76.5	80.1	77.2	541	704	52.1	-12
Tunisia	10,652	8,329	24.2	20.9	4.2	..	17.2	..	56	82	11.7	-8.7

(Continued)

Table 7.2 (Continued)

	Market Capitalization				Market liquidity[a]		Turnover ratio[b]		# domestic Co.[c]		S&P Global[d]	
	US$ Millions		% of GDP										
	2010	2018	2010	2018	2010	2018	2010	2018	2010	2018	2010	2018	
Turkey	302,443	149,264	39.2	19.5	52.2	48.2	133.3	248	263	377	21.4	-44
Ukraine	38,897	4,415	28.6	3.4	2.1	..	7.3	..	..	78	53.8	-17
United Arab Emirates	131,491	235,451	45.4	56.8	9.6	6.2	21.1	10.8	104	130	-6.8	-8
United States	17,283,452	30,436,313	115.3	148.5	240	161	208.4	109	4,279	4,397	12.8	-6.2
Vietnam	36,855	132,653	31.8	54.2	26.6	21.6	83.7	39.8	634	749	0.5	-6.1

Source: Prepared by the author based on World Bank data (World Bank, 2018).

Notes: This table demonstrates the evolution of stock market capitalization and the volume of shares negotiated around the world as a % of the GDP of each country. [a] Stocks traded, total value (% of GDP); [b] Value of domestic shares traded divided by their market capitalization; [c] # of Listed domestic companies; [d] S&P Global Equity measures the variation of the price of the dollar in markets covered by the indices S&P/IFCI and S&P/ Frontier BMI. These indices are intensely utilized by international portfolio managers.

Table 7.3 Structure of the ESG ratings adopted by the Fitch Ratings Agency

Environmental	Social	Governance
EAQ – Greenhouse Gas Emissions/Air Quality	SCR – Community Relations; Social Access & Affordability	GEX – Strategy Implementation, Operational Execution
EFM – Energy Management	SCW – Customer Welfare, Product Safety, Privacy & Data Security	GGV – Governance Structure
EWT – Water & Wastewater Management	SLB – Labor Relations & Practices	GST – Group Structure
EHZ – Waste & Hazardous Materials Management, Ecological Impacts	SEW – Employee Well-Being	GTR – Financial Transparency
EIM – Exposure to Environmental Impact	SIM – Exposure to Social Impacts	

Source: Prepared by the author based on information published by the Fitch Ratings Agency on its official website: www.fitchratings.com/site/esg (accessed on March 27, 2020). Each of the 14 items is evaluated on a 5-point scale as follows: 1 = Irrelevant to the entity rating and irrelevant to the sector; 2 = Irrelevant to the entity rating but relevant to the sector; 3 = Minimally relevant to rating, either very low impact or actively managed in a way that results in no impact on the entity rating; 4 = Relevant to rating, not a key rating driver but has an impact on the rating in combination with other factors; 5 = Highly relevant, a key rating driver that has a significant impact on the rating on an individual basis. Thus, according to the method adopted by the Fitch Agency: 1 or 2 = Lowest Relevance; 3 = Neutral; and 4 or 5 = Credit-relevant to the Issuer.

On the creditor side, some banks around the world allege that they are analyzing ESG risks closely to decide whether to loan money to certain corporations, even in emerging markets (Itaú, 2013). Seventy-seven percent of banks examine their loan portfolios in terms of ESG risks, according to the Fitch Agency. In some banks, this signifies assuming commitments to interrupt loans to companies in sectors seen as high risk in terms of ESG factors, such as those which manufacture firearms. In the view of some ratings agencies, the classification in terms of ESG factors requires greater diligence. However, from the point of view of rating agencies, currently banks rarely refuse to grant credit based just on ESG factors. However, banks in the ING Group (www.ing.com/) have adopted a procedure of granting loans at lower interest rates to companies that improve their ESG results. These loans, known as sustainability-linked loans, have increased in popularity in recent years. A current challenge that banks confront is developing quantitative models which estimate how, for example, the risks of global warming affect the financial performance of a loan. In terms of this, most banks are evaluating ESG risks qualitatively.

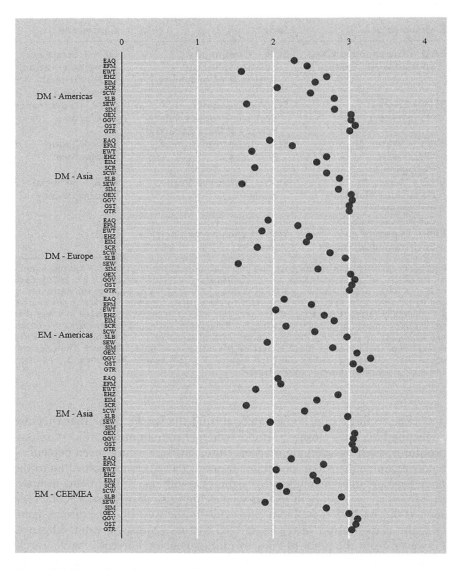

Figure 7.2 Typical performance of companies in terms of ESG according to the Fitch Ratings method (by region)

Source: Prepared by the author based on Fitch Ratings data.

Note: The legend for the vertical axis is defined in Table 7.3. N = 1,534 firms. DM = developed markets; EM = emerging markets; CEEMEA = Central & Eastern Europe, Middle East, Africa.

Figure 7.2 presents the values for ESG factors for 1,534 firms around the world, according to a method adopted by the Fitch Ratings Agency. The vertical axis consists of 14 items (listed in Table 7.3) which this ratings agency states that it considers to classify company risk in terms of ESG. Note that the governance items are those that receive the highest average values: GEX (Strategy Implementation, Operational Execution); GGV (Governance Structure); GST (Group Structure); GTR (Financial Transparency). A verification beyond the aggregate of the values attributed to companies permits a better understanding of this classification than average values.

For example, in environmental terms (E), Volkswagen and Oklahoma Gas & Electric Co. received ratings of 5 (Credit-Relevant to the Issuer) in EAQ terms (Greenhouse Gas Emissions/Air Quality), while Southern California Edison Company received ratings of 5 in EIM (Exposure to Environmental Impact). When observing social aspects (S), the Samarco Mineracao S.A. and Seplat Petroleum Development Company Plc appear with ratings of 5 in SCR (Community Relations; Social Access & Affordability). In social terms, Samarco appears with a rating of 5 in SIM (Exposure to Social Impacts), together with the JSC Medical Corporation EVEX, Seplat Petroleum Development Company Plc, Pacific Gas & Electric Company, and the PG&E Corporation. The Fitch Ratings Agency gave a 5 rating in terms of GTR (Financial Transparency) to: Samarco Mineracao S.A., Dalian Wanda Commercial Management Group Co., Ltd, Reward Science and Technology Industry Group Co., Ltd., Grupo Famsa, S.A.B. de C.V., Corporacion Electrica Nacional S.A. (CORPOELEC), and Petroleos de Venezuela S.A. (PDVSA). The governance item (G) that appears for the greatest number of sensitive companies is GGV (Governance Structure). Table 7.4 presents a group of 22 firms evaluated with ratings of 5 for this item according to the Fitch Ratings Agency.

In the rare occasions in which companies do not receive financing based on ESG risk factors, they generally have other financial options available. For example, when the banks in Australia reduced loans for new thermal coal projects and coal plants after the 2015 Paris Climate Accord, Chinese banks in response increased their loans to this sector according to a recent Fitch Ratings Agency report (Broughton, 2020).

Financial innovations, the capital market and sustainability

Just as the influenza pandemic a century ago taught us lessons which led to a new era for economies around the world, after the COVID-19 pandemic, consequences for the capital market appear to be inevitable. The occurrence of this pandemic should induce financial innovations within the context of the capital markets (Seyfang & Longhurst, 2013; Mendes-Da-Silva, 2019). Among the diverse lessons that can be learned from the COVID-19

Table 7.4 Issuers evaluated as very sensitive to governance structure in terms of credit (according to the method adopted by the Fitch Ratings Agency)

Issuer	Region	Sector
Panel A: Developed markets		
Volkswagen AG	Europe	Industrials & Transportation
Grupo Embotellador Atic, S.A.	Europe	Food, Beverage & Tobacco
Petropavlovsk plc	Europe	Natural Resources
CITGO Petroleum Corp.	Americas	Natural Resources
Pilgrim's Pride Corporation	Americas	Food, Beverage & Tobacco
Panel B: Emerging markets		
Albanesi S.A.	Americas	Utilities
Rio Energy S.A.	Americas	Utilities
Andrade Gutierrez Engenharia S.A.	Americas	Real Estate & Construction
Companhia Siderurgica Nacional (CSN)	Americas	Natural Resources
JBS S.A.	Americas	Food, Beverage & Tobacco
Odebrecht Engenharia e Construcao S.A.	Americas	Real Estate & Construction
Ouro Verde Locacao e Servico S.A	Americas	Generic
Automotores Gildemeister S.p.A.	Americas	Generic
Reward Science And Technology Industry Group Co., Ltd.	Asia	Food, Beverage & Tobacco
Global Cloud Xchange Limited	Asia	Services & Communications
Grupo Elektra, S.A.B. de C.V.	Americas	Retail, Consumer & Healthcare
Grupo Famsa, S.A.B. de C.V.	Americas	Retail, Consumer & Healthcare
TV Azteca, S.A.B. de C.V. (TV Azteca)	Americas	Services & Communications
JSC Holding Company United Confectioners	CEEMEA	Food, Beverage & Tobacco
Eskom Holdings SOC Ltd.	CEEMEA	Utilities
Corporacion Electrica Nacional S.A. (CORPOELEC)	Americas	Utilities
Petroleos de Venezuela S.A. (PDVSA)	Americas	Natural Resources

Source: Prepared by the author based on data published by the Fitch Ratings Agency (March 2020).

Note: This table presents the list of 22 (out of 1,534) companies whose capacity for credit is strongly conditioned on their corporate governance structure, according to the procedure adopted for ESG ratings adopted by the Fitch Ratings Agency. CEEMEA = Central & Eastern Europe, Middle East, Africa.

pandemic, the increase in Internet usage is something that has undeniably been observed.

In addition, the increase in financial instruments which offer protection to the wealth and jobs of people, to the financial health of companies, as well as to the resilience of countries, is something to be reflected on in the capital market. Thus, new products, new processes and new financial institutions should appear to preserve sustainability within the context of the capital market.

One lesson that it presents is that we should prepare ourselves for the economic and human consequences of the virus and act to minimize its impact. This pandemic represents a shock of supply and demand. Thus, it is expected that the economic crisis will be as contagious as the disease that has caused it. The work lost with the implementation of 14 days of self-isolation recommended for suspected cases will have serious economic implications. Closing regions and entire countries will cause a recession which could be more emphatic in regions whose markets are less resilient. It is within this scenario that financial innovations can contribute decisively to make quick and less expensive solutions viable to finance businesses, which in turn contribute to attenuating the effects of a recession due to COVID-19. The emergency reduction of interest rates in the United States and the United Kingdom should be the first of many policies designed to mitigate the economic impact of COVID-19.

New fiscal policy measures should come into play now as well. Individuals with poorly paid precarious work deserve specific attention. In cases in which medical care and medical leave are expensive, this can force people to work even if they are carrying the virus.

The appearance of financial innovations as a reaction to this growing demand through the observance of ESG factors points to a continual process of making efforts in this sense within the context of the capital market, with the involvement of traditional financial institutions, fintechs and large technological companies collaborating and competing with each other. In this sense, we've seen credit cards appear as well as recent financial products such as green funds (Chan & Tang, 2007), crowdfunding equity platforms (Baucus & Mitteness, 2016), green bonds (Flammer, 2018) and green stocks, among other innovations which may be on the way. In parallel to this, new companies will offer new financial services such as ESG ratings for companies listed on the stock exchange or for private equity.

But whatever these financial innovations are, it appears that there is a group of aspects that is common to most of them. In this respect, Eccles and Klimenko (2019) point to a sense of purpose on the part of companies, a promotion of engagement with investors, an increase in the involvement of the media in managing companies, investments in internal systems to manage the informational apparatus of ESG performance and the perfecting of metrics and reporting them directly to the market.

Financial innovations within the context of the capital market can assume forms and modes specific to needs and opportunities. For example, the real estate market has a growing demand for buildings judged to be more likely to protect the quality of life in cities. This occurs essentially through a group of factors which include, among other aspects, the intensive use of clean energy. And the fact is that buildings recognized as practitioners of positively viewed attitudes are preferred by renters who are willing to pay (and pay more) for this, even when we are dealing with less developed markets (Costa et al., 2018).

In other words, recent experiences point out that, if on one hand, industries such as agriculture and electricity are subject to changes in climatic variables, on the other, companies in other industries such as real estate can find opportunities. However, no matter which industry we're talking about, opportunities and threats are present, and investors will pay attention to the manner in which a company faces both sides of ESG aspects (Garcia et al., 2017).

At the same time that aspects related to sustainability gain relevance in society to the extent that they promote trust, resilience and the anticipation of risks in the context of the financial market, the finance literature has documented changes in the behavior of specific sectors of the market. For example, Crifo et al. (2014) study ESG factors in the financing of private equity. These authors conducted a field experiment involving 33 investors resulting in 330 observations. The results suggest that irresponsible ESG practices have a stronger impact than responsible ESG practices. In this sense, Crifo et al. (2014) find that irresponsible policies reduce a company's price by 11%, 10% and 15% in relation to E, S and G issues, respectively. On the other hand, responsible policies increase a company's price by 5%, 5.5% and 2% in relation to E, S, and G issues, respectively.

The bond market, in turn, is already exhibiting initiatives to explore opportunities for new products vis-a-vis ESG factors. In recent years, a large variety of companies, including Apple, Unilever and Bank of America, have produced green bonds to finance projects favorable to the climate. Despite this boom, little is known about the impact of these securities, but apparently they deliver environmentally positive results and also contribute to the financial performance of the issuing companies. According to Flammer (2018), a recent analysis of 217 green corporate securities emitted by companies listed on the stock exchange between 2013 and 2017 show that they produce a positive reaction from the stock market, improve financial and environmental performance, increase green innovations and increase the holding of these shares by long-term investors.

Concluding remarks

There remains no doubt that we are living through the beginning of a new era in terms of the relevance of sustainability issues in investor choices and preferences, and therefore in terms of the growth and development of capital markets. There is considerable space for conducting studies which provide clarifying evidence relating to the channels through which sustainable company practices can create value. There is undoubtedly a relationship between corporate practices and structures in accordance with ESG aspects, and this has an impact on the value of a company within the context of the capital market. This is based on the reaction of consumers, individuals and institutional investors, and even the imposition of restrictive rules on the part of governments.

However, research is necessary given that there already exists evidence which contradicts this expectation. For example, sin stocks tend to perform better than other stocks in the capital market. Companies that work with cigarettes, alcoholic beverages, gambling and other industries seem to contradict socially accepted rules and usually offer better performance than other companies, according to empirical evidence and discussions in the financial media (Mackintosh, 2019; Hong & Kacperczyk, 2009). In addition, there is news that banks currently are considering, but rarely rejecting, offering credit to companies based on environmental and social risks (Broughton, 2020).

Demands such as limiting future crises via global initiatives against climate change and the loss of biodiversity, for example, have become part of the daily agenda of each citizen, which is reflected by the capital markets vis-a-vis investor preferences. In addition, we have witnessed the establishment of behavioral norms for governments and international organizations. Up until now, the lack of standardization in the world of sustainable investments has not dissuaded investors from investing in this class of financial assets. Sustainability funds in the United States registered a record of US$21 billion in 2019, almost four times the amount invested in 2018 (Otani, 2020).

All of these issues taken together suggest that the reactions to sustainability aspects in the capital market are dependent on the extent to which players understand them as something that needs to be considered in order to preserve trust and resilience and anticipate risks. In this manner, governments, international institutions, companies and civil society itself should assume a more emphatically proactive role in this transformation, developing and supporting solid climate policies, green infrastructure projects and sustainable businesses in the broadest sense in accordance with ESG factors.

Note

1 In the beginning of 2019, the collapse of a Vale tailings dam in the city of Brumadinho, Minas Gerais, in Brazil, 80 miles away from the state capital Belo Horizonte, killed 270 people with a tsunami of dark-orange sludge, as the *Wall Street Journal* noted on December 31, 2019. According to the *Wall Street Journal*, based on interviews and documents, Vale's top management, including its CEO Mr. Schvartsman, paid scant attention to dams, which cost money and generated no return.

References

Abramov, A., Radygin, A., Entov, R., Chernova, M. (2017). State ownership and efficiency characteristics. *Russian Journal of Economics*, 3(2), 129–157. https://doi.org/10.1016/j.ruje.2017.06.002.

Albarrak, M.S., Elnahass, M., Salama, A. (2019). The effect of carbon dissemination on cost of equity. *Business Strategy and the Environment*, 28(6), 1179–1198. https://doi.org/10.1002/bse.2310.

Bachiller, P. (2017). A meta-analysis of the impact of privatization on firm performance. *Management Decision*, 55(1), 178–202. https://doi.org/10.1108/MD-12-2015-0557.

Baucus, M.S., Mitteness, C.R. (2016). Crowdfrauding: Avoiding Ponzi entrepreneurs when investing in new ventures. *Business Horizons*, 59, 37–50. https://doi.org/10.1016/j.bushor.2015.08.003.

Broughton, K. (2020). Banks taking a closer look at ESG risks in credit underwriting: Lenders are screening, but rarely rejecting, borrowers based on their environmental and social risks. *Wall Street Journal*, January 7.

CALPERS. (2014). *Towards Sustainable Investment and Operations: Making Progress – 2014 Report*. Available at: www.calpers.ca.gov/eip-docs/about/pubs/esg-report-2014.pdf.

Carlsson-Szlezak, P., Reeves, M., Swartz, P. (2020). What coronavirus could mean for the global economy. *Harvard Business Review*, March.

Chan, R., Tang, A. (2007). AIA-JF green fund – differentiation in funds market. *Harvard Cases*, #HKU664-PDF-ENG.

Cheng, B., Ioannou, I., Serafeim, G. (2014). Corporate social responsibility and access to finance. *Strategic Management Journal*, 35, 1–23. https://doi.org/10.1002/smj.2131.

Correia, S., Luck, S., Verner, E. (2020). Pandemics depress the economy, public health interventions do not: Evidence from the 1918 Flu. Available at SSRN: https://ssrn.com/abstract=3561560.

Costa, O., Fuerst, F., Spenser, J.R., Mendes-Da-Silva, W. (2018). Green label signals in an emerging real estate market. A case study of Sao Paulo, Brazil. *Journal of Cleaner Production*, 184, 660–770. https://doi.org/10.1016/j.jclepro.2018.02.281.

Crifo, P., Forget, V.D., Teyssier, S. (2014). The price of environmental, social and governance practice disclosure: An experiment with professional private equity investors. *Journal of Corporate Finance*, 30, 168–194. https://doi.org/10.1016/j.jcorpfin.2014.12.006.

Eccles, R.G. (2018). Why na activist hedge fund cares whether apple's devices are bad for kids. *Harvard Business Review*, January.

Eccles, R.G., Klimenko, S. (2019). Sustainability: The investir revolution. *Harvard Business Review*, May–June.

Flammer, C. (2018). Green bonds benefit companies, investors, and the planet. *Harvard Business Review*, #H04NXS-PDF-ENG.

Flores, E.S., Fasan, M., Mendes-Da-Silva, W., Sampaio, J.O. (2019). Integrated reporting and capital markets in an international setting: The role of financial analysts. *Business Strategy and the Environment*, 28(7), 1465–1480. https://doi.org/10.1002/bse.2378.

Garcia, A.S., Mendes-Da-Silva, W., Orsato, R.J. (2017). Sensitive industries produce better ESG performance: Evidence from emerging markets. *Journal of Cleaner Production*, 150, 135–147. https://doi.org/10.1016/j.jclepro.2017.02.180.

Garcia, L. (2020). Actis sees investors 'doubling down' on ESG after pandemic: A renewed focus on sustainability may boost investor appetite for socially useful infrastructure such as data centers. *Wall Street Journal*, June 3. Available at: www.wsj.com/articles/actis-sees-investors-doubling-down-on-esg-after-pandemic-1159 1138893?mod=searchresults&page=1&pos=3.

Green, J.P., Stark, A.W., Thomas, H.M. (1996). UK evidence on the market valuation of research and development expenditures. *Journal of Business & Accounting*, 23(2). https://doi.org/10.1111/j.1468-5957.1996.tb00906.x.

Hartzmark, S.M., Sussman, A.B. (2019). Do investors value sustainability? A natural experiment examining ranking and fund flows. *The Journal of Finance*, 74, 2789–2837. doi:10.1111/jofi.12841.

Hayat, U. (2015). ESG issues and investment practice. *CFA Institute Magazine*, 26(5), 50–50. https://doi.org/10.2469/cfm.v26.n5.17.

Hong, H., Kacperczyk, M. (2009). The price of sin: The effects of social norms on markets. *Journal of Financial Economics*, 93(1), 15–36. https://doi.org/10.1016/j.jfineco.2008.09.001.

Itaú. (2013). *ESG Integration into Fixed Income Research*. Available at: www.itau.com.br/_arquivosestaticos/Itau/PDF/Sustentabilidade/white-paper-esg-fixed-income.pdf. Acessed at March 28, 2020.

Kim, Y.B., An, H.T., Kim, J.D. (2015). The effect of carbon risk on the cost of equity capital. *Journal of Cleaner Production*, 93, 279–87. https://doi.org/10.1016/j.jclepro.2015.01.006.

Krosinsky, C. (2013). Sustainability and systemic risk: What's the SEC's role? *The Guardian*, October 23. Available at: www.theguardian.com/sustainable-business/sustainability-risk-investors-sec.

Mackintosh, J. (2019). Guns and coal offer the sin of tobacco stocks, but not the quality: Outperformance of alcohol and tobacco might have nothing to do with being sinful. The doesn't bode well for guns and coal. *Wall Street Journal*, December 3.

Mackintosh, J. (2020). Coronavirus will change the world. Good luck figuring out how: Our columnist contemplates five areas of uncertainty that leave the world on the edge of major changes. *Wall Street Journal*, May 28. Available at: www.wsj.com/articles/coronavirus-will-change-the-world-good-luck-figuring-out-how-11590686431?mod=searchresults&page=1&pos=5.

Mendes-Da-Silva, W. (2019). *Individual Behaviors and Technologies for Financial Innovations*. New York: Springer. ISBN: 978-3030063252. https://doi.org/10.1007/978-3-319-91911-9.

Mendes-Da-Silva, W., Onusic, L.M. (2014). Corporate e-disclosure determinants: Evidence from the Brazilian market. *International Journal of Disclosure and Governance*, 11, 54–73.

Mendes-Da-Silva, W., Onusic, L.M., Bergmann, D.R. (2014). The influence of e-disclosure on the ex-ante cost of capital of listed companies in Brazil. *Journal of Emerging Market Finance*, 13(3), 1–31. https://doi.org/10.1177/097265271 4550928.

Nauman, B., Gross, A. (2019). Credit rating agencies focus on rising green risks: Moody's warning on ExxonMobil marks watershed for debt markets. *Financial Times*, November 27.

Norton, L.P. (2020). 13 ESG investing trends to watch for in 2020 – Barrons.com: The outlook for environmental, social, and governance investing has brightened considerably. *Wall Street Journal*, January 30.

Orsato, R.J., Garcia, A., Mendes-Da-Silva, W., Simonetti, R., Monzoni, M. (2015). Sustainability indexes: Why join in? A study of the 'corporate sustainability index (ISE)' in Brazil. *Journal of Cleaner Production*, 96, 161–170. https://doi.org/10.1016/j.jclepro.2014.10.071.

Otani, A. (2020). Big technology stocks dominate ESG funds: The most commonly held S&P 500 stocks in actively managed sustainable equity funds last fall were giants including Microsoft, Alphabet and Apple. *Wall Street Journal*, February 11.

Pearson, S., Magalhaes, L., Kowsmann, P. (2019). Brazil's vale vowed 'never again'. Then another dam collapsed: The latest mining disaster took 270 lives and followed years of executive focus on trying to build the company into the largest in its industry. *Wall Street Journal*, December 31.

Roberto, M. A. (2011). BP and the gulf of Mexico oil spill. *Harvard Cases*, Product #W11366-PDF-ENG.

Seyfang, G., Longhurst, N. (2013). Growing green money? Mapping community currencies for sustainable development. *Ecological Economics*, 86, 65–77.

Watts, J. (2017). Long read: Operation car wash: Is this the biggest corruption scandal in history? *The Guardian*, June 1. Available at: www.theguardian.com/world/2017/jun/01/brazil-operation-car-wash-is-this-the-biggest-corruption-scandal-in-history.

World Bank. (2018). Indicators – World bank data – World bank group. Available at: https://www.worldbank.org/en/news/press-release/2019/10/29/world-bank-launches-sovereign-esg-data-portal.

Chapter 8

Sustainable alternate finance market

Anna Spoz, Piotr Niedzielski and Sylwia Henhappel

Summary

The first part of the chapter presents the concept of sustainable finance. Then, the structure of the alternative finance market is characterized, pointing out the importance of crowdfunding. Based on a critical analysis of the literature and the authors' own research, the relationship between crowdfunding and the concept of sustainable finance is presented. In the final section of this chapter, crowdfunding is discussed as an innovation.

The concept of sustainable finance

Most often, sustainable finance is defined as any form of finance that integrates environmental, social and corporate governance (ESG) criteria into the decision-making process in both the normal course of business and investment projects. The goal of this approach is to deliver lasting benefits to customers and society as a whole. The International Capital Market Association (ICMA) clarifies this definition to add that sustainable finance "incorporates climate, green and social finance while also adding wider considerations concerning the longer-term economic sustainability of the organisations that are being funded, as well as the role and stability of the overall financial system in which they operate." The scope of sustainable finance-related issues is given in Figure 8.1.

The term "climate finance" is understood as funding projects that mitigate the negative impact of the economy on the climate, for example, by reducing greenhouse gas emissions. "Green finance", on the other hand, is a broader concept, as it also covers funding environmental projects, such as the protection of natural resources, the conservation of biodiversity and the prevention and control of pollution. Further, "social finance" supports projects aimed at mitigating or resolving social issues and/or at generating positive social outcomes. Social finance project categories include but are not limited to the provision and/or promotion of affordable basic infrastructure, access to essential services (such as healthcare), affordable housing and

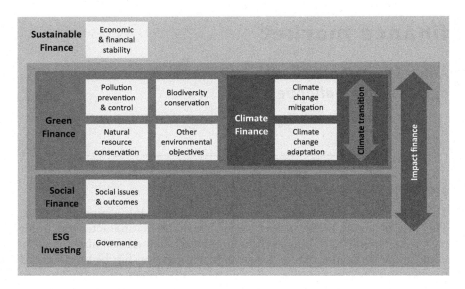

Figure. 8.1 The scope of sustainable finance–related issues

Source: Sustainable Finance: High-level definitions, International Capital Market Association, May 2020.

job creation, including through potential impact on small and medium-sized enterprises (SME) financing.

Alternative finance market size and structure

The essence of alternative financing is to raise and distribute funds excluding the traditional financial system (banking sector). The alternative finance market has seen a dynamic growth for several years now. This is especially conspicuous in the Asia-Pacific region (Figure 8.2). China has the largest market share in the region – over 71% (Ziegler et al., 2020, p. 35).

The United States has a dominant market share (96%) in the Americas; however, the Latin America and Caribbean market boasts a much larger increase. Compared to the Asian and American markets, the European alternative finance market is rather modest, but with stable growth.

In 2017, the size of the European alternative finance market was estimated at EUR 10.44 billion, with the largest share in the UK (68%). The other leaders were France, Germany and the Netherlands. The changes in the value of the alternative finance market in 2013–2017 demonstrate that although it grew almost 10-fold over that period, the year-on-year growth rate clearly decreases (Figure 8.3).

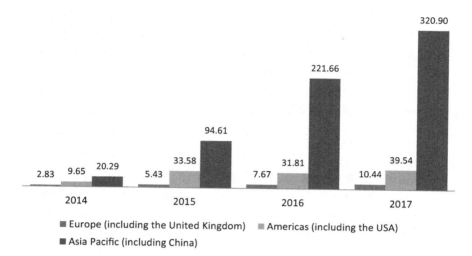

Figure 8.2 Regional online alternative finance market volumes 2013–2017 (EUR billion)

Source: Ziegler et al. (2019, p. 23).

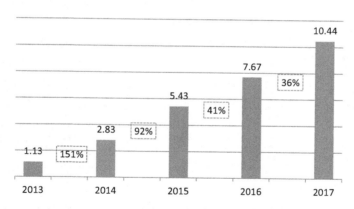

Figure 8.3 European online alternative finance market volumes 2013–2017 in EUR billion (including the UK)

Source: Ziegler et al. (2019, p. 22).

The UK alternative finance market in 2017 reached EUR 7.07 billion and is the highest in Europe. The second largest market in France was valued at EUR 661.37 million (i.e. only approximately 9.3% of the UK market). It is clear that the differences between countries with the largest and smallest alternative finance markets are enormous (Figure 8.4).

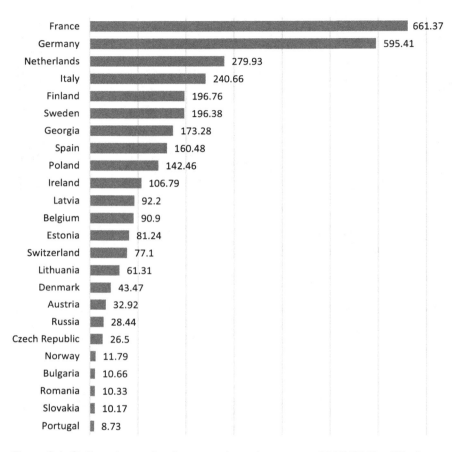

Figure 8.4 Online alternative finance volume by country 2017 (EUR million)
Source: Ziegler et al. (2019, p. 27).

The per-capita analysis demonstrates the importance of alternative finance for individual countries. The UK is a leader among European countries also in terms of this index, at EUR 107.04 per capita in 2017, but the countries ranking next are not the same as in Figure 8.4 (Figure 8.5). The lead looks interesting, with Estonia ranking second with EUR 61.76, and Latvia and Georgia ranking third and fourth, with EUR 47.51 and 46.62, respectively.

In terms of the number of online platforms that enable fundraising from alternative sources, the first three places in Europe are occupied by the same countries whose alternative finance market values are the highest: the UK – 77 platforms, France and Germany – 46 platforms each (Figure 8.6).

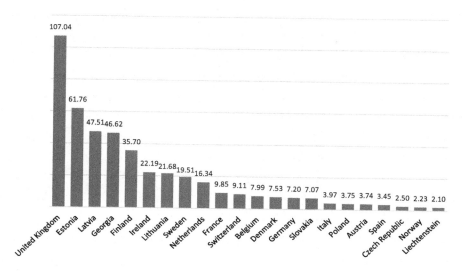

Figure 8.5 Market volume per capita by country in Europe 2017 (EUR)
Source: Ziegler et al. (2019, p. 29).

It should be noted that there are fewer locally based platforms in Germany than in France or Italy.

In 2018, the value of the alternative finance market in Poland totalled USD 333 million and sent the country up by two positions compared to the previous year in the ranking of European countries. Noteworthy, the total volume of finance for enterprises amounted to USD 41 million, the vast majority of which came from debt instruments (USD 34 million). Institutional investor volume reached USD 11.2 million (3%), the lowest share among key eurozone countries. The main driver of the social finance volume was peer-to-peer (P2P) consumer lending, accounting for 84.4% share (USD 281.4 million).

Crowdfunding and the concept of sustainable finance

Crowdfunding is a method of financing projects that are different in terms of types and sizes by the web community through various online platforms (Cumming et al., 2019). Crowdfunding is therefore an open invitation to provide financial support for a specific business project, venture or need, via an online platform. Consequently, crowdfunding can be interpreted as a specific service provided by specialized intermediaries (crowdfunding platforms) connecting the actors in the market game, i.e. "beneficiaries",

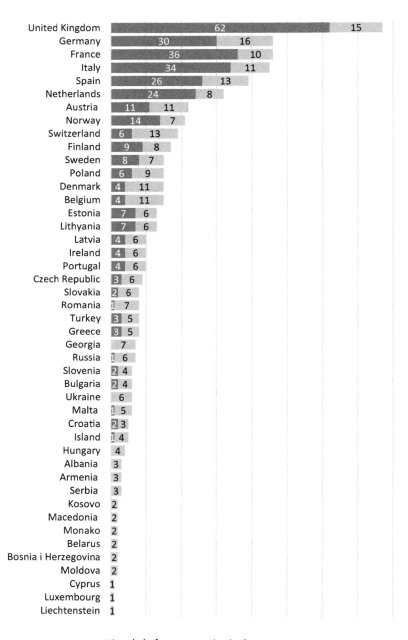

Figure 8.6 Number of online platforms on the alternative finance market in European countries 2017

Source: Ziegler et al. (2019, p. 25).

that are, creators or entrepreneurs ordering the crowdfunding service who want to raise funds (capital) for a purpose, project or venture with the other party, i.e. funders/donors, sponsors, lenders or investors (depending on the crowdfunding model). Electronic platforms that organize crowdfunding campaigns not only have to take actions to promote their solution or system among potential clients, both on the side of creators and potential donors, investors, lenders, i.e. on the side of crowdfunders, but also to play as intermediaries in the completion of transactions, present offerings for ventures, control their funding status, create interactions between market participants, offer new extra system functionalities, provide payment systems, manage accumulated funds and – depending on the crowdfunding model – reward participants and maintain system statistics (e.g. payments made, campaigns, number of participants). The actual range of these functions depends on the crowdfunding model, technologies used, system type or specificity.

This raises the question of how crowdfunding, and more broadly alternative finance, fits into the concept of sustainable finance. The authors will try to answer this question on the basis of a literature review and empirical research.

Martínez-Climent et al. (2019) completed a study on the relationship between crowdfunding and sustainable development. They analysed the number and type of publications and the most productive countries and journals in this regard. The study shows that since 2013 the number of publications with a link between "sustainability" and "crowdfunding" has been regularly growing. As many as 69% of these were scholarly articles. The most-cited review was Lama and Law in 2017, which was also indexed in the Green & Sustainable Science & Technology and Energy & Fuels categories. The others were assigned to such categories as Biotechnology & Applied Microbiology; Genetics & Heredity; Green & Sustainable Science & Technology; Energy & Fuels; and Environmental Sciences, which demonstrates that these publications deal with sustainable development, technology and energy development. The links between crowdfunding and environmental issues are also clear from the titles of the journals that featured the papers published. The most have been published in the *Journal of Cleaner Production, Sustainability* and *Renewable Sustainable Energy Reviews*.

Messeni Petruzzelli et al. (2019) pointed out that a crowdfunding campaign is more likely to be successful in the case of sustainable development–oriented projects than commercial ventures. The success of crowdfunding campaigns focused on sustainable development can significantly contribute to the promotion of behaviours in line with sustainable society models among citizens. Bento et al. (2019) arrived at similar conclusions, showing that crowdfunding is an attractive, creative source of funding for sustainable development projects. They demonstrated that highlighting the underlying sustainable character of a project helps the campaign to succeed. They also showed that the average survival rate (above 70%) after one year of operation proves the rationale behind the origination of healthy projects for sustainable development.

The importance of crowdfunding platforms in enhancing and facilitating the diffusion of social and cultural projects through donations was noticed by Presenza et al. (2019).

Butticè et al. (2018) examined how the institutional environment impacts the spread of green crowdfunding campaigns in various countries. Based on an "original machine-learning algorithm", they analysed 48,598 campaigns announced on the Kickstarter platform between 1 July 2009 and 1 July 2012. The analysis shows that the probability that a crowdfunding campaign is "green" is lower in those countries where institutions are oriented towards sustainable development, due to easier access to funding for green projects from traditional sources. Crowdfunding is therefore an alternative source of financing for green projects, helping the originators mitigate restrictions in access to capital.

Van der Have and Rubacalba (2016) argued that crowdfunding platforms, through the development of new products and services, respond to social needs that are not satisfactorily met by the market or the public sector. Crowdfunding platforms contribute to overcoming societal challenges by creating social relationships and a collaboration model. Langley et al. (2020) indicated that crowdfunding facilitates social entrepreneurship in Berlin and enables solidarity economies.

Crowdfunding can be used as an alternative financing option for corporate social responsibility (CSR) projects. For enterprises, the use of a crowdfunding platform to pursue CSR projects not only allows them to source extra funds for their implementation but also to gather and engage stakeholders around these initiatives. Individuals and social groups supporting a shared project or cause develop ties with each other, have a sense of shared responsibility and remain loyal to the project owner. This process of co-creation and participation improves the empowerment and engagement of stakeholders, ultimately to offer an enterprise the opportunity to expand its market reach. Based on her study, Witoszek-Kubicka (2020) demonstrated that it is possible for enterprises to use crowdfunding portals as vehicles for CSR initiatives through patronage of a project, support of a collection, organisation of a collection or support of the platform itself.

Taking the concept of sustainable finance as project finance (for current and investment projects) integrating ESG criteria into the business or investment decisions, we examined the goals and nature of crowdfunding campaigns published on online platforms. It is obvious that for a campaign to be successful, both its goal and the adopted method of communicating with potential investors/donors must be not only acceptable to them but also gain their interest and sympathy (Gera and Kaur, 2018; Bagheri et al., 2019). It should be noted at this point that the attractiveness of a project to the "crowd" does not necessarily mean its usefulness for the society, environment or the way an entity operates.

A study of the most popular Polish crowdfunding platforms showed that the majority of crowdfunding projects are social and some of them also environmental; however, no governance projects were found, which by nature is quite plausible (Table 8.1).

Crowdfunding statistics published by Massolution in 2017 reveal that nearly 19% of projects were socially related, which is a significant share in global crowdfunding market.

The authors also analysed the ten largest crowdfunding platforms and found that all of them host socially and environmentally oriented projects or projects that can have impact on the environment or communities. Some platforms have separate categories for such projects, but in fact, projects of this kind can be found in any category, e.g. arts, food, healthcare, wellness, education, technology. There are also whole platforms dedicated to environmental and social projects. Table 8.2 presents the most popular categories of crowdfunding projects and shows which platform supports these categories. The first ten platforms are the largest ones, and the last four are platforms that are best for investing in ecology, renewable energy, communities and sustainability.

The issue of effective and efficient sourcing of funds for projects has been extremely relevant for years. The alternative finance market (private equity, business angels or crowdfunding) is a mean to address these problems. Private equity and venture capital funds enable the reduction of legal and capital barriers in the finance sourcing process, yet they entail a higher risk. Business angels (BAs) are another, non-institutional form for finance sourcing, which consists in the provision of capital support for a planned

Table 8.1 Nature of crowdfunding campaigns on the most popular Polish online platforms 2013–2019

Crowdfunding platform name	Current campaigns			Successfully completed campaigns		
	Social	Environmental	Governance	Social	Environmental	Governance
Polak potrafi [A Pole can do it]	1	1	-	26	10	-
Wspieram To [I support it]	-	1	-	14	6	-
Wspólnicy [Partners]	0	0	-	no data	no data	-
Zrzutka [Fundraiser]	19	2	-	25	5	-

Source: Authors' own study.

Table 8.2 Crowdfunding platforms and categories with a high share of social and environmental projects (marked with +)

Platform	Technology	Green tech/Clean tech	Social/Community	Creative/Art	Sports	Fashion/Textiles	Food & Beverage	Theater & Dance	Education	Health/Wellness	Environment/Ecology	Energy
1 Kickstarter	+			+		+	+	+				
2 Indiegogo	+	+	+	+		+	+	+	+	+	+	+
3 GoFundMe			+	+	+					+		
4 Crowdfunder	+	+		+		+	+		+	+		+
5 Fundly			+		+				+	+		
6 Seedrs						+	+	+		+		+
7 Start Engine	+	+	+	+	+	+	+		+	+	+	
8 Fuel a Dream	+		+		+		+	+	+	+	+	
9 Kiss Kiss Bank Bank			+	+	+	+		+			+	
10 Crowdcube		+			+		+			+		
11 Goteo	+		+						+		+	
12 Ecocrowd			+			+	+		+		+	+
13 Citizen energy											+	
14 Triodos Bank			+								+	+

Source: Authors' own study.

project by wealthy and experienced entrepreneurs. Unfortunately, this form of financing is used rather rarely, is operative for low-value projects and usually entails the originator's loss of the title to most of the profits.

Crowdfunding (especially social lending and investment crowdfunding) is an interesting attempt to finance micro and small-sized economic projects, including start-ups. A major advantage of this form of financing is lower regulatory restrictions, with the main limitation, on the other hand, being the maximum amount that can be sourced, which in Poland was set at EUR 1 million.

The size and dynamics of the development of crowdfunding make it difficult to ignore or downplay it. The European Commission has faced a dilemma: whether to put in place EU-wide regulations, thus protecting users and ensuring the inviolability of the crowdfunding idea, or, for fear that regulations would slow down or inhibit the growth of this instrument, allow it to develop freely. So far, at the European Union level, there is no single legal regulation to govern the rules of crowdfunding. A number of member states of the European Union have developed national regulations in this regard (e.g. Belgium, Germany, the Netherlands), and preparations to do so are underway in some others. The number of cross-border transactions means that solutions should be sought at the international level (Figure 8.7).

Discrepancies between the provisions of national and EU law make it difficult for operators to conduct cross-border crowdfunding activities. The lack of a single regulation for crowdfunding at the European Union level, combined with the simultaneous introduction of differing regulations nationally, lets the barriers to the cross-border expansion of crowdfunding increase instead of decrease. In a draft regulation, the European Commission proposed to introduce the institution of an authorisation to operate as a provider of crowdfunding. According to Article 10 of the draft regulation, the authorisation is to be granted by European Securities and Markets Authority (ESMA), and the finance provider status will be allowed to incorporate bodies that meet the requirements for the protection of potential investors.

Crowdfunding platforms will have the option to decide whether they want to operate under the current rules (i.e. in accordance with national regulations) or apply for the authorisation that will allow them to operate throughout the European Union, without the risk of having to submit to any additional requirements.

For now, the proposed solution is to apply only to equity and debt crowdfunding, in which funding offers do not exceed EUR 1 million, calculated for a period of 12 months for a specific project. This means that authorised platforms will only be able to organise campaigns to a limited extent. When

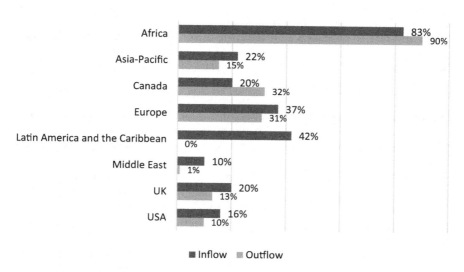

Figure 8.7 Inflow and outflow rate of cross-border transactions by region (2018)
Source: The Global Alternative Finance Market Benchmarking Report. Cambridge, 2020.

implementing projects with a higher level of funding, it will be necessary to meet more stringent requirements under other legal acts, such as the new prospectus regulation.

Under the regulation, providers will have to ensure compliance with conflict-of-interest rules. According to Article 7, crowdfunding platforms will not be able to run campaigns in which they would hold any financial participation. A similar prohibition will apply to campaigns originated by platform shareholders holding 20% or more of the share capital or voting rights, members of their management or employees. Providers will be required to adopt appropriate procedures to identify and disclose any conflicts of interest that should be communicated to potential investors.

Crowdfunding as innovation

The growth of crowdfunding is one of the examples of modern innovation processes. The interest in innovative developments in the services sector or the leveraging of innovation to boost competitiveness can be considered at this point as well. The approach to innovation has changed throughout history, and one can assume that this is a process directly correlated with the directions of economic development; the frequent modifications found in theory and in practice make this field of knowledge highly creative.

The core factor in the growth of this form of fundraising and funding business projects or other ventures of various natures is the development of the global Internet network and the rising importance of knowledge as a special-property resource in the modern economic processes. At the same time, crowdfunding, as a kind of financial intermediation service, fits well in the innovative processes characteristic of this sector. Therefore, when discussing service innovation in the form of crowdfunding, one should bear in mind innovation as such and only complement it with the specific nature of the services sector that influences the shaping of innovation. Innovation can thus be a new service, a new way of providing a service and/or a new way of organizing a service business. Electronic platforms provide crowdfunding services in their role of market brokers and create an electronic space for integration of the other groups, i.e. funders and beneficiaries, within the system. At the same time, the practice assumes a relatively large – and potentially unlimited – number of participants (i.e. donors, sponsors, investors, lenders, beneficiaries, etc.) and their free access to the system (network). In some systems and platform operation models (especially investment platforms), the number of participants is often very limited in practice, e.g. only to accredited investors.

For one to speak of crowdfunding as a service innovation, one must determine whether the new service, the new way of providing the service and/ or the new way of organization meets the requirement of innovation and whether it will bring measurable economic and social benefits, e.g. in the form of easier access to funding for social and business projects. Service innovation cannot be limited to a change in the characteristics of the service itself. It is often associated with the introduction of new ways of distribution, interaction with the client, quality control, security, etc. In view of the specific nature of the services and innovative processes sector, Pim Den Hertog distinguished four dimensions of innovative behaviour in services:

- new service concept,
- new client interface,
- new service delivery system,
- new technological options.

These aspects and the relationships between them, as well as underlying conditions, are presented in Figure 8.8.

As indicated, the key dimension in the development of crowdfunding is the use of new technologies, i.e. the Internet and network development, e.g. in the form of Web 2.0. Financial intermediation, with crowdfunding being a type of such financial service, can be assessed and described in the three dimensions indicated in Figure 8.8, i.e. a new service concept (Dimension 1), a new client interface (Dimension 2) and a new service delivery system (Dimension 3).

A **new service concept** is, in other words, the provision of a new type of service. The provision of funding for various types of projects is a need of

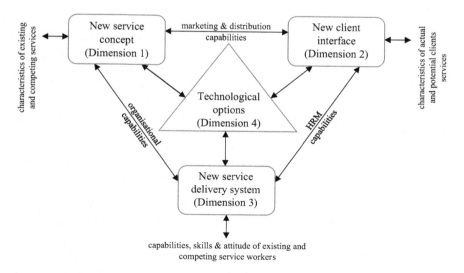

Figure 8.8 Dimensions of service innovation

Source: Hertog (2000, p. 19).

the modern economy. The opportunities brought by the development of the Internet have also been recognized in the area of financial intermediation and the development of banking services. The new concept refers to the fact that crowdfunding platforms enable accumulation of distributed capital donors, or funders. In the classic model, the party had to be an entity (a bank or other institution, e.g. a stock exchange, a brokerage firm), which through its operations was able to organize or provide such funding while assuming most of the risk associated with financing a business idea, with the risk borne indirectly, e.g. by clients of the bank who had deposited their money in it. In the case of a stock exchange, the risk was directly transferred to those buying shares. In the crowdfunding platform model, the risk is transferred directly to capital donors in proportion to their financial contribution, while entities/persons with low financial strength can be participants on the capital supply side (microlending). The concept of financing through crowdfunding uses the mechanism of interactive added value creation, defined by Reichwald and Piller (2009, p. 51).

A new client interface means the introduction of a new way of cooperation not only of the crowdfunding platform but also of the parties to the crowdfunding practice. The development of crowdfunding services indicates that the role of this method of fundraising will dynamically develop and change. Not without significance is the activity of the crowdfunding platforms themselves, which will seek to attract customers, i.e. attracting and serving ideas to be funded, as well as the Internet community ready to financially support (in various models) these ideas. As the trends of the crowdfunding industry itself show, projects related to the broadly defined concept of sustainable growth, including pro-ecological projects, increasingly often find financing there. It is a result of the possibilities for reaching with an idea out to a broad (practically unlimited) audience, who are at the same time a potential source of funding at minimized transaction costs. Very often, creators are young people with some industry experience, who are very good at navigating through and using online solutions, for whom the priority is not to get funding for their idea but in the first place to promote a certain idea or solve a specific problem using advanced web-based tools. Not without significance is the fact that those willing to invest according to their preferences and values, who for various reasons do not accept the offer of traditional banks, also seek alternative capital investment opportunities that enable them to support certain values or ideas at the same time.

A new delivery system for financial intermediation services via crowdfunding platforms uses automation and algorithmization to reduce the costs of transactions made on the crowdfunding platform. Simultaneously, a partial specialization of platforms can be observed. A number of crowdfunding models are described in the literature,[1] which is also related to the specialization and the applied technological solutions of individual crowdfunding platforms. One can indicate a division of crowdfunding into investment, lending, reward and donation-based models. For the purposes of this chapter, crowdfunding can generally be divided into investment and non-investment categories, with financial investment crowdfunding being an area of special interest from the point of view of the opportunities and role of crowdsourcing in sustainable socio-economic growth and sustainable finance (Figure 8.9).

The use of new technological options is associated with the conquest of the Internet landscape by technologies such as Web 2.0, 3.0 and even 4.0. Technical advances in the field of information and communication technologies, as well as the rapid spread and development of the Internet, constantly open opportunities for communication, access to all kinds of information and electronic services, as well as organization of work and spare time. Such mobile devices as notebooks, smartphones or tablets or online tools, e.g. social media or other advanced technologies, including cloud computing, as well as the "network effect" itself are commonplace and only provide a few examples of the proliferation of information and communication technologies related to the development of the information society. The progressing digitization makes the real world merge increasingly with the virtual and increasingly gets mirrored in the network. Technical and technological progress has had a huge impact on all aspects of human life and activity, but today progressive technical solutions are the driving force of modern, constantly developing societies, which use the opportunities offered by participation in the global online community almost in an unlimited way.

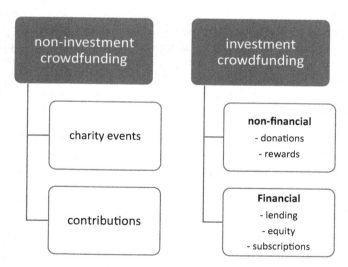

Figure 8.9 Crowdfunding taxonomy
Source: Authors' own study.

Companies that recognize significant changes in their operating environment in the right time have a better chance to adapt and ensure good future by, for example, instilling new intelligence in existing structures. According to Professor B. DeLong of the University of California, Berkeley, "Information technologies and the Internet are just as powerful in strengthening the mind as the inventions of the industrial revolution used to be in increasing muscle strength".

Since the beginning of the 21st century, an abundance of web-based applications has been created that are grouped under such keywords as social network, community applications, online community, etc. All these keywords share one feature: the resulting collective social intelligence, i.e. support for social networks, communities and collective transfer of knowledge on the Internet through appropriate, globally available applications and often also relatively easy-to-use technologies. Collective intelligence is also getting more and more attractive for companies that want to make decisions on a large scale, engage their clients in innovation processes and solve problems by user action. Just as with community development on the Internet, users have more influence over decision-making and promoted trends. According to Schwartz, this is the "age share" that can weaken or even replace central control over management approaches. In addition to the growing importance of user impact, one can notice that in online community applications a large number of users often produce equivalent or even better solutions than those presented by centralized or expert knowledge-based approaches. This is also the case with the **market opportunity assessment** for projects that seek alternative/innovative funding sources, which crowdfunding certainly is. So, to some extent, the concept of crowdfunding uses the phenomenon also known as "wisdom of the crowd", fitting well into the wider practices of crowdsourcing. This new business approach model provides applications that concentrate and share the knowledge of a wide range of users. The community takes over tasks such as unsolved problems, including problems of fundraising or finding more effective ways to invest financial capital, where a large group of users is either better or cheaper compared to computers or experts.

Projects based on the crowdfunding concepts are increasingly being used by public administration bodies, e.g. high level of transparency in the budget decision process (Lee et al., 2016). In this sense, according to Davenport et al. (2016), crowdfunding is similar to open innovation. Within this approach, he shows that organizations that want to survive in the market in the future need a variety of "strategies for acquiring key knowledge through employees". The participation of clients in the innovation process of enterprises using Web 2.0 or later is, among others, a new form of knowledge acquisition strategy, such as self-help communities as part of enterprise discussion forums, or alternative sources of funding for the development of or, more broadly, shaping of a competitive advantage in which also the user, and not just the enterprise, solves the problems of others (Nucciarelli et al., 2017). *In the open model* a multidirectional flow of ideas is assumed between the enterprise and its environment. This approach provides for a combination of external and internal ideas into systems that meet the requirements of business models. A business model uses internal and external resources to create value and defines internal mechanisms that allow the company to consume part of this value. In turn, internal ideas can be commercialized through external channels, outside the company's current business, in order to generate extra value.

So, using the Internet today is not just about consuming information. Today, via the Internet, people have the opportunity in the global arena to quickly and interactively participate in various tasks, including in fundraising, finding alternative ways to invest their financial capital or a specific assessment of an idea and its improvement. People are today often actively involved in creating content and even added value. This may be googling to fill memory gaps or fight ignorance in a field, creating knowledge in Wikipedia, sharing information with family or friends on Facebook, blogging reviews from trips in exchange for holiday postcards and solving complicated tasks or participating in research and development projects, as well as financing various types of projects. With the network, people from different parts of the world are able to simultaneously, i.e. without temporal or geographical limits, interact with each other, which is the starting point for crowdsourcing, one of the forms of which crowdfunding in all its varieties is. Thus, crowdfunding as an example of crowdsourcing falls under the open innovation model.[2] The paradigm of open innovation models was pointed out by Chesbrough (2006) as currently trending. Moreover, the analysis shows that these models in the current conditions of the economy are more favourable in terms of efficiency than closed models. The basic difference in the pursuit of innovative processes in open versus closed models is the opening of the process (at various stages) towards the outside of the enterprise, which is associated with acquiring the knowledge necessary to pursue innovation from the external environment and making use of it, as well as the commercialization of valuable knowledge or technology – not directly related to the company's business – including in the form of spin-off companies.

Based on the global spread of the Internet and information technology, a certain phenomenon has been observed, called crowdsourcing by J. Howe in 2006 in *Wired* magazine. He describes this phenomenon as follows:

Technological advances in everything from product design software to digital video cameras are breaking down the cost barriers that once separated amateurs from professionals. Hobbyists, part-timers, and dabblers suddenly have a market for their efforts, as smart companies in industries as disparate as pharmaceuticals and television discover ways

Business projects	Subscriptions, small lending **Subscribers**	Innovative projects **Visionaries**
Social projects	Public collections **Social activists**	Petty savers **Wizards**
	Form of reward – community (symbolic, minor)	Form of reward – business type (loans, profit share, equity share)

Figure 8.10 The BCG-type taxonomy matrix for crowdfunding projects
Source: Authors' own study.

to tap the latent talent of the crowd. The labor is not always free, but it costs a lot less than paying traditional employees. It's not outsourcing; it's crowdsourcing.

However, for enterprises, to implement the crowdsourcing approach in their own business is a great challenge due to the combination of personal individual benefits of employees with the benefits of the organization.

One promising strategy is the integration of personal information management for the benefit of enterprise knowledge management, i.e. the transfer of personal information to enterprise knowledge. Another challenge is to integrate collective decisions into the company's strategy. To summarize the discussed issues, the Boston Consulting Group (BCG)-type taxonomy matrix for crowdfunding projects is proposed (Figure 8.10).

Conclusions

Crowdfunding gives the opportunities to get financing for projects that for various reasons are not able to raise funds from other sources. Obtaining financing requires convincing the backers that a project is worth undertaking. That is why crowdfunding campaigns appeal to backers' emotions and values (Robiady et al., 2020). To answer the question whether crowdfunding is a method of raising money for sustainable projects, the results of scientific research were analysed. We also studied categories of projects announced on the ten most famous crowdfunding platforms. It has been shown that each of the crowdfunding platforms give opportunities to launch campaigns in accordance with sustainable goals, and the number of publications in this scope is systematically growing.

Beside broadly analysing the financial aspect of crowdfunding, attention should be paid to the non-financial advantages of crowdfunding, which include but are not limited to options to raise funds without losing ownership title in the project and/or participation in profits, carry out marketing research for interest in the product/service among potential clients, diagnose their needs at a stage when it is still possible to modify the product/service (Hu et al., 2015; Brown et al., 2017; Viotto da Cruz, 2018) and conduct a promotional campaign. Thanks to these advantages, publishing a project on a crowdfunding platform gives its creators the opportunity to confront their ideas about the needs and expectations of clients and their idea of a new product or service against reality (Stanko and Henard, 2017). The campaign uniquely combines the market research process with a promotional campaign of a product or service among potential buyers (Li et al., 2019).

The growth of the crowdfunding industry carries certain risks. The most important ones are related to the publication of an idea on an online platform and thus making the idea public at the design or prototype stage without any guarantee that the intended goal will be achieved, underestimation of crowdfunding costs, credibility of operating crowdfunding platforms operators and their privileged position in relation to entities registering their projects.

One of the most serious risks associated with registering a project on an online platform and making it available to a wide group of Internet users is the fear that the proposed solution may be copied, so it is very important at the stage of preparing a crowdfunding campaign to obtain legal advice on intellectual property rights protection and discuss these issues with the administrator of the selected online platform.

One should also be aware of the moral hazard resulting from information asymmetry (Belleflamme et al., 2015). In the case of crowdfunding, donors and investors do not have control over the direction of spending funds raised in the campaign. This may lead to a situation where the transferred money will be spent contrary to its original purpose or projects will be executed without using financial discipline, i.e. due care for the efficiency in executed tasks. According to Strausz (2017), a way to reduce moral hazard is to defer the payment of funds obtained, i.e. until the end of the collection, and not to disclose the size of the entity raising the funds.

There has also emerged an accusation in the opinions of crowdfunding scholars that crowdfunding distorts competition, as the costs of finance for similar projects for similar entities may vary. Such differences may result from both the adopted method of project finance (traditional, e.g. loan, and crowdfunding) and the operating rules of specific online platforms. In response to these concerns, in December 2019, after seven years of work, the European Parliament came to an agreement with the Council of the EU on the harmonization of the principles for the operation of crowdfunding platforms in the EU. In practice, this means that crowdfunding for economic operators (lending-based and equity-based crowdfunding) will be covered in the near future by EU-level regulation in a harmonized framework.

Crowdfunding is an example of an innovative process that allows projects to be carried out in line with the concept of sustainable finance (Messeni Petruzzelli et al., 2018). Crowdfunding platforms, by reaching a wide audience, can contribute to an increase in the awareness of this concept among society.

Notes

1 For more on taxonomies see: Spoz (2016).
2 In terms of research and development effort (R&D), the literature distinguishes basically between closed innovation models and open innovation models. For more see: Niedzielski (2013).

References

Bagheri, A., Chitsazan, H., Ebrahimi, A. (2019). Crowdfunding motivations: A focus on donors' perspectives. *Technological Forecasting and Social Change*, 146, pp. 218–232.

Belleflamme, P., Omrani N., Peitz M. (2015). The economics of crowdfunding platforms. *Information Economics and Policy*, 33, pp. 11–28.

Bento, N., Gianfrate, G., Thoni, M.H. (2019). Crowdfunding for sustainability ventures. *Journal of Cleaner Production*, 237, 117751.

Brown, T.E., Boon, E., Pitt, L.F. (2017). Seeking funding in order to sell: Crowdfunding as a marketing tool. *Business Horizons*, 60 (2), pp. 189–195.

Butticè, V., Colombo, M.G., Fumagalli, E., Orsenigo, C. (2018). Green oriented crowdfunding campaigns: Their characteristics and diffusion in different institutional settings. *Technological Forecasting and Social Change*, 141, pp. 85–97, 2019.

Chesbrough, H.W. (2006). *Open Innovation: The New Imperative for Creating and Profiting from Technology*. Boston: Harvard Business Press.

Cumming, D., Johan, S., Zhang, Y. (2019). The role of due diligence in crowdfunding platforms. *Journal of Banking & Finance*. doi:10.1016/j.jbankfin.2019.105661.

Davenport, S., Daellenbach, U., Leitch, S. (2016). Extending theorisation for open innovation and absorptive capacity via a social capital lens. *ISPIM Innovation Symposium*. Manchester: The International Society for Professional Innovation Management.

Gera, J., Kaur, H. (2018) A novel framework to improve the performance of crowdfunding platforms. *ICT Express*, 4 (2), pp. 55–62.

Hertog, P.D. (2000). Knowledge-intensive business services as co-producers of inno-vation. *International Journal of Innovation Management, World Scientific Pub-lishing Co. Pte. Ltd.*, 4 (4), pp. 491–528.

Howe, J. (2006). The rise of crowdsourcing. *Wired*, 14 (6). [online]. Available at: www.wired.com/2006/06/crowds/ [Accessed 31 March 2020].

Hu, M., Li, X., Shi, M. (2015). Product and pricing decisions in crowdfunding. *Marketing Science*, 34 (3), pp. 331–345.

Langley, P., Lewis, S., McFarlane, C., Painter, J., Vradis, A. (2020). Crowdfunding cities: Social entrepreneurship, speculation and solidarity in Berlin. *Geoforum*, 115, pp. 11–20. doi:10.1016/j.geoforum.2020.06.014.

Lee, C.H., Zhao, J.L., Hassna, G. (2016). Government-incentivized crowdfunding for one-belt, one-road enterprises: Design and research issues. *Financial Innova-tion*, 2 (1). doi:10.1186/s40854-016-0022-0.

Li, Y.-M., Liou, J.-H., Li, Y.-W. (2019). A social recommendation approach for reward-based crowdfunding campaigns. *Information & Management*. doi:10.1016/j.im.2019.103246.

Martínez-Climent, C., Costa-Climent, R., Oghazi, P. (2019). Sustainable financing through crowdfunding. *Sustainability*, 11 (3), p. 934. doi:10.3390/su11030934.

Messeni Petruzzelli, A., Natalicchio, A., Panniello, U., Roma, P. (2018). Understand-ing the crowdfunding phenomenon and its implications for sustainability. *Techno-logical Forecasting and Social Change*, 141, pp. 138–148, 2019.

Niedzielski, P. (2013). *Kreatywność i procesy innowacyjne na rynkach usług trans-portowych. Ujęcie modelowe.* Szczecin: Polskie Towarzystwo Ekonomiczne, pp. 70–93.

Nucciarelli, A., Li, F., Fernandes, K.J., Goumagias, N., Cabras, I., Devlin, S., Kudenko, D., Cowling, P. (2017). From value chains to technological platforms: The effects of crowdfunding in the digital game industry. *Journal of Business Research*, 78, pp. 341–352.

Presenza, A., Abbate, T., Cesaroni, F., Appio, F.P. (2019). Enacting social crowd-funding business ecosystems: The case of the platform Meridonare. *Technological Forecasting and Social Change*. doi:10.1016/j.techfore.2019.03.001.

Reichwald, R., Piller, F. (2009). *Interaktive Wertschöpfung: Open Innovation, Individualisierung und neue Formen der Arbeitsteilung.* Wiesbaden, Germany: Springer Gabler, 2nd ed., p. 51.

Robiady, N.D., Windasari, N.A., Nita, A. (2020). Customer engagement in online social crowdfunding: The influence of storytelling technique on donation perfor-mance. *International Journal of Research in Marketing.* doi:10.1016/j.ijresmar.2020.03.001.

Spoz, A. (2016). Crowdfunding – nowinka czy nowy perspektywiczny model finan-sowania projektów biznesowych? *Roczniki Ekonomii i Zarządzania*, 8 (44), pp. 184–204.

Stanko, M.A., Henard, D.H. (2017). Toward a better understanding of crowd-funding, openness and the consequences for innovation. *Research Policy*, 46 (4), pp. 784–798.

Strausz, R. (2017). A theory of crowdfunding: A mechanism design approach with demand uncertainty and moral hazard. *American Economic Review*, 107 (6), pp. 1430–1476.

Sustainable Finance: High-level definitions. (2020). International Capital Market Association, May. [online]. Available at: www.icmagroup.org/assets/documents/Regulatory/Green-Bonds/Sustainable-Finance-High-Level-Definitions-May-2020-110520v4.pdf.

The Global Alternative Finance Market Benchmarking Report. (2020). Cambridge: Trends, Opportunities and Challenges for Lending.

Van der Have, R.P., Rubacalba, L. (2016). Social innovation research: An emerging area of innovation studies? *Research Policy*, 45 (9), pp. 1923–1935.

Viotto da Cruz, J. (2018). Beyond financing: Crowdfunding as an informational mechanism. *Journal of Business Venturing*, 33(3), pp. 371–393. doi:10.1016/j.jbusvent.2018.02.001.

Witoszek-Kubicka, A. (2020). Wykorzystanie platform crowdfundingowych do realizacji działań w ramach społecznej odpowiedzialności biznesu. *Social Inequalities and Economic Growth*, 61.

Ziegler, T., Shneor, R., Wenzlaff, K., Odorovic, A., Johanson, D., Hao, R., Ryll, L. (2019). *Shifting paradigms.* The 4th European alternative finance benchmarking report. Cambridge: The Cambridge Centre for Alternative Finance.

Ziegler, T., Shneor, R., Wenzlaff, K., Wanxin, B., Jaesik, K.W., Odorovic, A., Ferri de Camargo Paes, F., Suresh, K., Zheng Zhang, B., Johanson, D., Lopez, C., Mammadova, L., Adams, N., Luo, D. (2020). The global alternative finance market benchmarking report. Trends, opportunities and challenges for lending, equity and non-investment alternative finance models. *The Cambridge Centre for Alternative Finance*, p. 35.

Chapter 9

Sustainable rating agencies

Daniel Cash

Introduction

In the aftermath of the financial crisis of 2007/8, the very nature of the financial system, along with its effectiveness, viability, and social utility was questioned (Walker et al. 2016, p. 17). A number of progressive responses were advanced, but one that has taken a particular hold is the concept of *sustainable investing*, or at least the 'mainstreaming' of a concept related to sustainable investing. The rate of development in this field of sustainable investment since the crisis has been rapid, with major coordinated efforts (like that of the UN-sponsored Principles for Responsible Investment initiative) seeking to advance the doctrine to the major parts of the financial arena. However, the traditional concepts of investing 'responsibly' or 'sustainably' have moved on since before the crisis so much that traditional modes of investing such as ethical or impact investing are now being overshadowed by the need to 'blend' such sentiments with economic return on investment as well (Mendell and Barbosa 2012, p. 111). This 'mainstreaming' of the concept of sustainable investment contains a number of core principles. One of those, as Pagano et al. inform us, is that 'every investment needs a benchmark' (2018, p. 339). Furthermore, 'in a competitive market economy, all economic actors are constantly calibrating their performance in relative and absolute terms against their peers' (ibid). So, in a post-crisis era containing market participants that want to succeed yet now, hypothetically, understand more the importance of considering factors other than outright financial performance, how may those market participants know which investment is right for them to make? One way that risk can be reduced is by utilising the calculated opinions of sector experts. For that reason, understanding *sustainability rating agencies* is important.

To the uninitiated, the concept of a 'sustainability rating agency' is likely one that can be deciphered rather easily: an agency that exists to rate an entity (or product) on how it incorporates non-financial information in its processes, thereby making it more or less 'sustainable'. Yet the moment one enters the literature in this related field, it quickly becomes a terminological minefield that contains so many different definitions and meanings for invented terminologies that it is sometimes difficult to understand the true lay of the land. There are instances where agencies are called agencies, but they are merely indexes containing information of market participants. There are other instances where those who develop *rankings* are regarded as rating agencies, which then feeds into academic research which perpetuates the terminological minefield and makes an area of the marketplace complicated, when there is little need to do so; there is enough complication in the financial sector to go around. So, in that vein, this chapter seeks to provide a 'lay of the land' for the area of *rating agencies* who actually *specialise* in the rating of non-financial factors that may affect an economic entity or product. That is not to say that indices, rankings, and the more-established credit rating agencies who have started to incorporate large portions of the smaller sustainable rating market will not be mentioned. They are all an important part of the overall framework and, in certain cases, may actually provide insight into the potential trajectory of the sustainable rating industry.

Yet, for this chapter, the concept of a *sustainable rating agency* is the one that we will be focusing upon mainly. To do that, the chapter will contain a number of areas which will help us achieve the goal of obtaining a solid understanding of this niche-but-growing industry. First, it will be important to incorporate context for the development of the industry. To accentuate this understanding, we will need to know more of what has been happening since the financial crisis, since the apparent move to a more sustainable approach to conducting financial business is now starting to fundamentally dictate the development of these associated industries that seek to serve it. Once that assessment is completed, we will introduce the identified sustainable rating agencies. From this point on I will use the acronym 'SRA' to refer to sustainable rating agencies, taking great care not to overburden with the use of acronyms which are awash in the literature. This will help differentiate from the many other factors in this area, which we will also need to be introduced to. However, we shall see that this terminology is itself troublesome, even when simplified.

Learning about the development of the industry will be useful, but there are issues that are seemingly inherent within this industry. This is common in industries that seek to serve the larger financial marketplace (like credit rating agencies or auditors). Yet to better understand those fundamental issues we will need to know more about the methodologies that the agencies are applying when determining how sustainable a given entity is. Therefore, the third section of the chapter will analyse the methodologies of the SRAs in direct and abstract detail. This will lead us into a discussion regarding the issues impacting the development of the industry and, more crucially, the impact that they may have upon the trajectory of the industry. The SRA industry is under a particular threat in that it provides something useful but is in the realm of much larger and more resourceful competitors, namely the leading credit rating agencies. I have made the argument that the two industries will likely not coexist for long elsewhere, and understanding the issues affecting the SRA industry as it progresses may feed into that argument as it develops (Cash 2018).

Understanding more the potential future of the industry is very important. This is because

> ESG ratings and indexes are a crucial component of the way business is done in the twenty-first century because they ensure key business issues are reflected in company assessments and tracked by relevant benchmarks and investment indexes, and ultimately become actionable to investors and other end users.
>
> (Pagano et al. 2018, p. 339)

The move towards a more progressive and forward-thinking financial community has been an important and, arguably, vital reaction to the financial crisis. That SRAs play a part in that movement is extremely positive. However, are they actually suited, or at the very least, truly capable of providing the support that movement needs? What would the provision of that support look like if the credit rating agencies began to devour more of the smaller SRA industry? These abstract questions will be addressed throughout and, as is the objective of the chapter, should become more answerable once the chapter concludes.

Why do we need sustainable rating agencies?

The concept of investing for anything other than financial recompense is a long and storied one. However, the concept of 'sustainable investing' has its roots in the so-called 'Brundtland Report' of the 1987 World Commission on Environment and Development's Report on the issues of finance and sustainability (World Commission on Environment and Development 1990). As Weber comments:

> Sustainable finance is finance that meets the social, environmental, and livelihood needs of the present generation without compromising the ability of future generations to meet their own needs and that creates a fair balance between societies in the north and the south.
>
> (Weber 2015, p. 121)

It is this connection between present and future generations that lies at the heart of the concept, and unsurprisingly why it has gained momentum since the financial crisis that witnessed a 'massive wealth transfer' (Bhagat 2017,

p. 23). However, it was well before the financial crisis that these ideas were advanced; it just took the crisis to spur a change in mentality, it would seem. Other concepts were advanced which have begun to take hold, like the concept of the 'triple-bottom-line', which takes environmental, social, and governance (ESG) factors into account *equally* with economic issues. These foundational concepts were the basis for the Brundtland Report, which stressed that finance had an *active* role to play in the development of a sustainable system.

Writing after the financial crisis, Miles suggests that, at its very broadest, the concept of sustainable finance refers to the 'mainstreaming of environmental and socio-economic criteria into lending, investment and other financial services' (Miles 2013, p. 240). Walker et al. agree, noting that in the reaction to the crisis, a large number of interested parties, ranging from practitioners to scholars, 'have explored a variety of ways in which the markets can be structurally changed to avoid excessive risk taking, the widespread focus on short-term performance, self-serving behaviour, and outright fraud' (Walker et al. 2016, p. 17). Furthermore, Hebb discusses the development that institutional investors are beginning to realise that they have a greater role to play in the oversight of the financial system. She talks about how, rather than the institutional investor 'socialism' that Peter Drucker imagined, it is the case that we are witnessing a reconfiguring of the concept of capitalism itself (Hebb 2011, p. 1). Hebb suggests that this is based upon the fact the institutional investors are beginning to engage more in the running of the companies they are a part of, although she does contest that this is because, since the crisis, 'investors are increasingly concerned about risk in their portfolios' (ibid 2). There are a number of connotations to this sentiment. The first is that the financial crisis was, for a large part, a failure in internal governance. A number of entities have been investigated since the crisis and were found to be partaking in massively risky business without any capability to minimise that risk, or in some cases simply outright fraud. Second, more and more research is being developed that suggests that those who consider factors other than just purely economic performance stand a better chance of success in the long run (ibid 4). This is known as the 'business case' for responsible investing. It is also aligned to the reality that in the current system dominated by extraordinarily large institutional investors, it is difficult for an institutional investor to just walk away from their position (known as the 'Wall Street Walk' according to Hebb) because they are so widely invested; so, rather than walking, the modern approach is to engage and alter the performance of the company in question.

This move to a more sustainable system has been labelled a 'megatrend' (Lubin and Esty 2011, p. 2). This megatrend is supposedly based upon the fact that externalities, such as carbon dioxide emissions and water usage, are quickly becoming 'material' in the minds of investors (amongst a host of other reasons). However, the shift to this new mode of thinking is not finished and has certainly not been a smooth transition so far. Robins argues that despite the current financial elite having their roots in the post-WWII welfare era that produced the large pension and mutual fund industries, the process of 'financialisation' has somewhat corrupted that foundational sentiment. He refers to a number of studies which have shown that hundreds of billions of dollars have been 'diverted from shareholders and customers to investment banking employees alone over the past two decades' (Robins 2012, p. 5). Additionally, he argues, the time horizon in financial thinking has continued to *decrease* via the development of associated technologies and industries (like the hedge-fund industry), when what is called for in the sustainable sense is an *increase* in that time horizon. Yet more foundational or philosophical barriers need to be crossed in order for the sustainable vision to truly take hold.

The concept of investing responsibly traditionally involved the application of 'screens'. These screens would be applied according to the investors' tastes and appetites. For example, one may choose to invest but not in anything connected to the arms industry. These screens could be positive or negative, so rather than not investing in something because it was connected to the arms industry, one may choose to invest in vehicles that produced positive externalities for a particular community, for example. The specialised thought took hold that incorporating non-financial issues into investment decisions could be advantageous economically, with sustainable investors being the first to raise climate change as a financial issue. Robins muses whether the issue for traditional sustainable investors was 'being right too early' because the 'mainstream' was particularly dismissive at first and is only now playing catch-up (ibid 6). However, it is not the case that classical financial theory has recoiled and retreated. Rather, conventional financial theory is still somewhat dismissive of the impact of non-financial factors. It has been argued that the assumption from conventional theory is that 'the introduction of non-financial factors will harm diversification and thereby incur penalties in terms of risks and returns' (ibid 11). Yet there is an ever-increasing amount of data that suggest investments selected on the basis of identifiable ESG factors 'do tend to outperform' investments that ignore such data (ibid). This was also found to be the case when it came to fund performance, with sustainable-focused funds outperforming those that focused on ethical aspects only. The overarching sentiment being advanced by proponents of this model is simple: 'investors need to understand, measure and promote superior financial *and* non-financial performance' (ibid 12).

That may well be the case, and it is also true that the concept is gradually maturing (Trinks and Scholtens 2017, p. 193), but there are issues still to be resolved. Philosophically, the transition to the 'mainstream' is proving to be a particularly relevant issue. The suggestion has been that the transition has led to a 'suspension of its critical edge'. Also, there is the issue that with the new focus being on making sure the sustainable focus in economically relative, so as to go 'mainstream', responsible investors are now only focusing on ESG issues to the extent that they are 'material', which many have argued goes against the concept of investing responsibly (Robins 2012, p. 15). This may also lead to a lessening of massively important issues, like human rights. Butz and Laville label this the 'materiality gap', and it is something that will plague the transition of the concept (Butz and Laville 2007). It has been identified that the most important 'battleground' for advocates and opponents has been the concept of financial performance (Krosinsky 2012, p. 19), with Revelli stating that socially responsible investing has become 'financialised' (2017, p. 711). Revelli continues by discussing the development of the concept of 'ESG integration', which involves the consideration of ESG factors by mainstream investment managers of 'some' key ESG criteria, i.e. material ESG data. However, the theoretical backdrop to this incorporation is stark, with conventional financial wisdom – Revelli identifies a source of this wisdom as emanating from the Chicago School, where Friedman openly criticised corporate social responsibility (CSR) as a 'constraint on profit maximisation in a capitalist economy' – promoting the importance of financial considerations above all else. This has led others to wonder whether the adoption of sustainable practices in the investment arena 'was merely market rhetoric rather than a sign of serious commitment to social responsibility in business and finance' (Hellsten and Mallin 2006, p. 403). It has also been argued that this modified form is being held out as the utilitarian vision of CSR, and hence why it is being adopted by the mainstream.

If there are philosophical issues with the modern version of sustainable investment, there are certainly technical issues too. The first is that the entire field is awash with different terminologies, often for the very same thing. It has been suggested by Krosinsky and Robins, probably accurately, that this is a natural result of people trying to define an ever-changing and evolving sector. As the scholars say, 'one woman's "ethical investing" is another man's "socially responsible investing", and one firm's "responsible investing" is another manager's "sustainable investing" . . . in such a fluid field, we are all well aware of the dangers of false precision' (Krosinsky and Robins 2008, p. xxi). Yet the changeability, however understandable, is problematic. It can lead to a level of disengagement, as classical responsible investors have their sector affected by the mainstreaming of the concept, which is based upon very different foundations, as we have seen. In line with this, the lack of a definition means that, as Cullis et al. argue, 'the notion of sustainability is even more vague [than CSR] and all-embracing' (2015, p. 605).

If we remove the definitional issues from the discussion for one moment, the question of how this new form of data is even quantified, collected, and disseminated is also an important one. Hill helpfully notes that there are a number of concerns surrounding the quality of the data being reported regarding the sustainability of financial entities, including the fact that some are forced to declare via regulations, whilst others are not; a uniformity of relevance across sectors cannot be possible; a lack of coverage with only 5% of the world's public companies reporting their own ESG-related issues, like emissions-related data, for example; and many others (Hill 2020, p. 175). We shall return to these issues of data and its availability later in the chapter, as it is important when understanding the potential efficiency and usefulness of the SRAs. However, it has been widely noted that an increase in both the availability and reliability of ESG data is required if the ideal of a sustainable system is to be achieved (Țîrcă et al. 2019, p. 10); a number of regions, including the EU and the United States, are currently implementing legislation aimed at rapidly improving the rates of ESG-related disclosure (Rezaee 2019, p. 183).

Yet there is one outstanding question that arguably dictates all of the discussions here; simply, what do mainstream investors actually want? Hedstrom commented recently that the answer can be seen in their actions. In 2017, Vanguard joined their competitors BlackRock and State Street, the world's largest institutional investors, in voting *against management* in moves which were seen as watershed moments (voting actions against the management at ExxonMobil and Occidental Petroleum in particular). Hedstrom notes that, in 2018, more than 400 shareholder resolutions were filed based on ESG issues. Eight resolutions were put forward *demanding* that the boards of companies address climate risk issues in a major way, whilst 70 were put forward in a similar manner regarding the equal treatment of women. There have been massive divestments from industries related to fossil fuel usage (Hedstrom 2018, p. 68). Hedstrom suggests that these actions are telling. They may well be, particularly as such actions have only increased since the time of Hedstrom's writing. However, there are some glaring issues that can be seen if we are to suggest that companies and investors need to do more with regard to sustainability. How would an investor know if a company was truly sustainable in its practices? How would investors know how one company compares to another when it comes to their consideration of ESG-related factors in their operations? With there being issues regarding the availability of ESG-related data from within companies, how can the investor base get closer to those companies in a way which is palatable to the issuing companies but which does not create extra costs for the investor? Arguably, the answer to all those questions and more can be found in the concept of the *sustainable rating agency*.

The sustainable rating picture

The need for third-party verification within the capital markets, in a number of different realms, is a constant. Dispersed investors need information that they can understand and incorporate easily. Those same investors may seek to constrain the actions of their agents by way of an independent third-party verifying tool. Managers may seek to demonstrate their actions to their investors and may need that demonstration to be supported by an entity the investors deem independent from the management. Issuing companies may need to 'signal' to the marketplace their credibility or their operational processes. They may also need to do this without actually revealing what may be commercially sensitive information. These sentences constitute what has been advanced before as 'signalling theory' in the rating arena (Cash 2018) and, in reality, I could easily have removed the word 'may' from all of the sentences. I could have done this because all of these elements, within the confines of the credit markets, are truths. In analysing the dynamics of the credit markets, we see that there is a fundamental need for third-party verification. In the credit rating industry, the post-crisis regulatory approach to reduce the foothold of the leading 'Big Three' rating agencies has not only failed but witnessed the credit rating agencies becoming *more* established; it is because of this fundamental truth that the need to *signal* within the dynamic is fundamental.

So, as the mainstreaming of 'sustainability' continues, it stands to reason that the role of signaller would grow with it as well. This has been the case, with the associated third-party verifying industries witnessing tremendous rates of growth alongside the development of sustainability. However, as was mentioned in the introduction, the field from the view of the literature is a definitional nightmare, with a number of entities being regarded as 'a', when it fact they really ought to belong to 'b', and so on. In this section of the chapter we will be introduced to the leading SRAs, alongside the other players within the field who will be categorised accordingly. As an example, the definition provided by Diez-Cañamero et al. provide us with a good place to start.

The authors attempt to provide a review of the entire marketplace which is particularly needed. The article starts with an accurate summation, in that:

> The financial market is pushing the development of Socially Responsible Investment, which has led to the rise of Corporate Sustainability Systems. These CSSs are tools that rate corporate performance on sustainability. However, they constitute a chaotic universe, with instruments of different nature. This paper identifies and groups the common characteristics of the CSSs into three different typologies: Indexes, Rankings, and Ratings.
>
> (Diez-Cañamero et al. 2020, p. 2153)

This starting point provided by the scholars is particularly helpful. The differentiation into the three typologies provides for clear marks of demarcation. To do this they utilise the *Rate the Raters Report 2019* from SustainAbility (2019), which in surveying market participants on the usefulness of ESG rating agencies, declares that the following entities are 'agencies' as we would clarify them:

- CDP Climate, Water & Forests Scores
- RobecoSAM (and its Corporate Sustainability Assessment)
- Sustainalytics (its ESG Risk Ratings)
- MSCI (ESG Ratings)
- Bloomberg (ESG Disclosure Scores)
- ISS-Oekom (Corporate Ratings)
- FTSE Russell (ESG Ratings)
- ISS (QualityScore)
- EcoVadis (CSR Rating)
- Thomson Reuters (ESG Scores)
- Vigeo-Eiris (Sustainability Rating)

This is a mostly accurate representation, although some modifications need to be made to improve its accuracy, as it is not definitive and should not be relied upon as such. Even in this 'definitive' list, a number of entities are not 'agencies' in the classic sense, which may lead to confusion. Furthermore, agencies such as Standard Ethics are excluded, with no reason provided. Also, the reference to 'Thomson Reuters' needs to be updated to consider recent Merger & Acquisition (M&A) activity, as well as some other issues. So, to that end, we shall be introduced to entities earlier, in turn, whilst attempting to add to the list for completeness.

CDP climate, water, and forest scores

The first entity on the list is the CDP's Climate, Water, and Forest Scores. Founded as the Carbon Disclosure Project in 2002, the global non-profit organisation is comprised of the CDP Worldwide Group and CDP North America, which are controlled by a board of trustees and a board of directors, respectively. The organisation scores companies and cities on their disclosure of relevant issues relating to environmental transparency and action and, as of 2019, have had 8,400 companies and 920 cities disclose through their systems. Each entity is rated according to the climate change/water/forest themes, with aspects such as a focus on reducing carbon emissions, water usage, and links to supply chains that affect deforestation all considered. The ratings scale runs from A to F, with + and – modifiers. As of 2017, the organisation developed its 'A-List' system, where 200 corporate entities and 105 cities were rated as highly ranked entities but had their rankings diversified by a certain sector; for example, Danone was one of the very few that achieved 'A' in all three sectors, whilst Ford only scored 'A' in Climate and Water, not Forests. What is very interesting indeed is that according to SustainAbility's *Rate the Raters* report, the CDP was cited as the most useful by market participants. This is interesting because, as Shahzad et al. note, 'the weakness of the data is that reporting is voluntary which may bias data' (2019, p. 552). Additionally, the data are relied upon as is and are not verified by external means. We will return to this in the next section, as it is a particular issue affecting certain parts of the ESG rating sector.

RobecoSAM

In 1995 Reto Ringger founded SAM (Sustainable Asset Management), which was one of the first investment companies focused on ESG as a concept. In 2006, the much more established Robeco company – founded in 1929 in Rotterdam – acquired a 64% stake in SAM, and the two merged in order to establish a stronger position in the burgeoning ESG marketplace. Having progressed within the market with its 'Corporate Sustainability Assessment' (CSA), the ESG Ratings division of the company was 'transferred' to S&P Global (one of the big three credit rating agencies) in 2019 and would maintain its identity as SAM under the agreement, with Robeco maintaining access to its data. S&P launched its 'S&P Global ESG Scores' product, which fully incorporates SAM's product, in May 2020.

Every year, before the transfer to S&P, RobecoSAM would invite 2,500 of the world's largest firms by float-adjusted market capitalisation according to the S&P Global Broad Market Index to participate in the annual CSA. The CSA is designed to identify and capture both general and industry-specific criteria regarding ESG. On average, each theme contains six to ten criteria, which equates to between 80 and 120 questions for companies to answer. Each criterion is worth up to 100 points, and each is assigned a specific weight against the total of the questionnaire. Then, for each company, 'a Total Sustainability Score' of up to 100 points is calculated based on the predefined weights established for each question and criterion. Companies receive a Total Sustainability Score between 0 and 100 and are ranked against other companies in their industry (Pagano et al. 2018, p. 350). As opposed to the CDP, 'to ensure the accuracy of companies' self-reporting, RobecoSAM takes a number of measures, including cross-checking companies' answers with publicly available information and using third-party audit of the process conducted by Deloitte (ibid). This is important, because the highest-ranked companies are then selected for inclusion in the Dow Jones Sustainability Indices (DJSI). As part of S&P's purchasing of SAM, the CSA has now been rebranded as the 'SAM Sustainability Yearbook', although it is exactly the same process.

Sustainalytics

Sustainalytics is a mixing of a number of companies – Jantzi Research, DSR, Siri Company, and GES International – which all came together by 2009. In 2010 it launched its Country ESG Risk Research & Ratings service to accompany its corporate ratings service. In 2011, STOXX launched the Global ESG Leaders ESG Index, which relied upon data delivered to it by Sustainalytics (Caré 2018, p. 145). Since then, Sustainalytics has partnered with Bloomberg to have its resources shared via Bloomberg terminals and with ISS to have its research inserted into the proxy voting process. Then, in 2017, the credit rating agency Morningstar acquired a 40% ownership stake in Sustainalytics in an attempt to keep pace with the ever-expanding big three credit rating agencies. In 2020, Morningstar acquired the remaining 60% of the agency to become the sole owner.

The agency's rating scale is from 1 to 100. Its methodology is split via assessments of ESG impacts and, depending on industry, specific weights will be attached to certain criteria. It has at least 70 indicators for each industry and, in relation to its assessment of ESG impact, such assessments are divided between preparedness (what systems and policies are in place to help manage ESG risks); disclosure (does the company meet international best practice standards); and performance (both qualitative and quantitative). Before an ESG Rating Report is published by the agency, a draft report is sent to the company being rated. The aim is to gather more information, if useful, and gather feedback (Huber and Comstock 2017).

MSCI ESG Ratings

MSCI ESG Ratings is a clear example of a *service*, rather than a stand-alone *agency*. The ESG ratings service stands within the large MSCI company, standing for Morgan Stanley Capital International. The ESG ratings service was launched in 2010, and despite the fact that it sits within another company, it is one of the largest providers of ESG ratings; it provides ESG ratings for more than 6,000 companies globally and more than 400,000 equity and fixed-income securities. Its rating scale runs from AAA to CCC, and it looks at 37 key ESG issues, divided into the three pillars of ESG. The predominant themes assessed are climate change; natural resources; pollution and waste; environmental opportunities; human capital; product liability; stakeholder opposition; social opportunities; corporate governance; and corporate behaviour. The data for these ratings are collected from a range of sources, including governmental databases, company disclosures, and macro-data made available from academic, governmental, and non-governmental organisation (NGO) databases. Before a rating is published, companies are invited into a formal data verification process.

Bloomberg (ESG scores)

If the MSCI ESG Scores are one of the largest ESG ratings providers, then Bloomberg's ESG Disclosure Scores are one of the most widely available sets. This is because, like the MSCI ESG Scores, Bloomberg's offering sits within the larger Bloomberg company and are thus available via all of its 'terminals'. It provides at least some data on more than 100,000 companies globally (Shahzad et al. 2019, p. 551), utilising CSR and/or sustainability reports, annual reports and websites, and direct contact with companies. They cover 120 ESG factors, including carbon emissions, supply chains, resource depletion, discrimination, diversity, and shareholder rights. The data are checked and standardised and, interestingly, they will penalise companies for 'missing data'.

ISS-Oekom

Immediately, this 'agency' needs to be clarified. In March 2018, the proxy-voting service ISS (Institutional Shareholder Services), which was owned by MSCI until 2014 and is now part of Genstar Capital, acquired Oekom Research AG and incorporated it into its organisation; ISS-Oekom ratings was formed. However, very quickly the company rebranded its offering so that it now stands as ISS ESG. ISS used to provide its own rating on 'G'-related issues under the moniker of ISS QualityScore, but this is now simply just one component of a list of other components that are being powered by its acquisition of Oekom; they now provide ESG rating services in the fields of corporate, country, disclosure, sustainability bonds, carbon risk, and custom ratings. The now-titled Governance QualityScore itself contains in-depth research on nearly 6,000 publicly traded companies globally. Its rating scale runs from the first to tenth decile, with those rated in the tenth decile constituting a higher risk. It analyses over 100 criteria for each section and utilises both publicly available information and dialogue with the rated companies to arrive at its final decision. Across all of the different fields of ratings, the firm uses a 'prime' status identifier for those who score highly.

FTSE Russell ESG Ratings

FTSE Russell's ESG Ratings service provides data on more than 7,000 securities across nearly 50 developed and emerging markets. It provides research and ratings for investors which are built on the usual division of the three pillars of ESG and then more than 300 individual indicators across the three pillars. Interestingly, the company is clear in their communication that 'ratings are calculated using an exposure-weighted average, meaning that the most material ESG issues are given the most weight when determining a company's score'; this is interesting, semantically if nothing else, because the usage of the term 'material' is not universally witnessed across the providers.

EcoVadis

EcoVadis started in Paris in 2007 and has a distinct selling point – its focus is on helping procurement teams and the supply chain. In 2008 it started producing 'scorecards' and, in the same year, joined the United Nations Global Compact. It declares that it is the world's 'most trusted provider of business sustainability ratings, intelligence and collaborative performance improvement tools for global supply chains'. In terms of its development, it recently received a $200 million investment from CVC in January 2020, and a month later received an undisclosed investment from Bain & Co., which signalled the trajectory of its development. It has more than 600 employees and 60,000 rated companies. Its methodology is founded on seven principles: ratings must be *evidence based*; the *industry sector, country, and size matter*; there must be a *diversification of sources*; *technology is a must*; the ratings must be conducted by *international CSR experts*; there is always *traceability and transparency*; and everything must be founded upon the underlying principle that one can achieve *excellence through continuous improvement*. The rating is split across four different pillars as opposed to the usual three ESG pillars, and they are environment, labour and human rights, ethics, and sustainable procurement. In order to verify the data, a number of avenues for information are followed, including publicly available information, information from the rated company, and external third-party endorsements. Within each of the four pillars, the company is scored according to a specific adjusted weighting criteria – usually 25% for policies, 40% for actions, and 35% for results, which is then weighted against the particular industry and the size of the company – with a total score of 100 available. The agency will chase up non-responsive suppliers in order to complete its assessments, as it requires 'documentation from suppliers to support answers and data provided' (Bateman et al. 2016, p. 139). Bateman et al. note the usage of EcoVadis for the global supply chain, as 'buyers have an organisation collecting the necessary data for them, and suppliers have a repository for their information from which to draw for other requesting buyers' (ibid). It has also been noted that services such as those provided by EcoVadis may be needed for compliance purposes, for example, when seeking to prove a product conforms to legislation (perhaps regarding the use of harmful substances, for example) (Johnsen et al. 2014, p. 187).

Thomson Reuters ESG Scores

Thompson Reuters is another example of an offering that needs to be clarified. In 2009 Thomson Reuters purchased Asset4 and, in 2015, established the Thomson Reuters ESG Scores. However, in 2018 as part of its joint creation of *Refinitiv* with BlackRock, the vast majority of Thomson Reuters's financial and risk unit elements were transferred to the new entity, including its ESG ratings capability – the transferring valued the new entity at around $20 billion. How that will progress is continuing to unfold, but the Asset4 methodology will continue to be central to its ESG rating processes. To further emphasise the ever-changing nature of this industry, at the time of writing the London Stock Exchange is awaiting approval from EU antitrust regulators to press ahead with their $27 billion acquisition of Refinitiv.

Writing in 2016, Herriott notes how Asset4 ESG contained data on nearly 5,000 public companies worldwide. It did not apply preliminary screening, and its methodology breaks down the usual three pillars of ESG into subcategories that are derived from 226 key performance indicators (2016, p. 29); indicators include human rights, vision and strategy, emission reduction policies and practices, employment quality, and board structure, amongst others. The firms rating is determined by the comparison between its raw values against others in a comparison group, i.e. 'E' scores are compared against other 'E' scores within a particular industry, and so on. The scores can also be compared geographically (usually the 'S' pillar) or can be compared universally (sometimes this applies to 'G' and/or 'E'). The indicator scores, calculated to have a range of 0 to 1 based on minima and maxima in the comparison group, are then rolled up into subcategory scores and are then weighted according to a number of attributes, including relevance to the comparison group (more weight added); wide reporting within the comparison group (more weight added, with break points at 5% and 15% of the group); the distribution of raw scores (less weight to narrowly distributed indicators); correlation with other indicators (less weight to highly correlated ones); types of indicator (policy variables get less weight than activity variables, which get less than outcome variables); and finally consistency with academic literature (more weight). Finally, each of the three pillars are weighted equally in the total sustainability score for a firm (ibid).

Vigeo-Eiris

Vigeo-Eiris has been noted as being one of the earliest SRAs, if not the earliest (Berg et al. 2019, p. 5). It is, however, important to accurately chart its development. It is actually the EIRIS component that represents the tradition of the firm, with EIRIS being founded in 1983; Vigeo was founded in 2002. EIRIS was originally a British charity set up to aid churches and charities with incorporating ethical perspectives into their investment decisions. In 2001 EIRIS partnered with FTSE to launch the FTSE4Good series. Vigeo, after acquiring a number of smaller rating entities on the European continent, joined forces with EIRIS in 2015 to become Vigeo-Eiris. In 2019, the credit rating giant Moody's purchased Vigeo-Eiris to bolster its ESG offerings.

Its methodology for rating divides 38 precise sustainability criteria across six domains. These are then segmented into 41 sector sub-frameworks that is where relevant weights are applied via sector and geography, etc. These then run through 330 indicators. The larger framework is pinned to international standards (like the UN, the OECD, or the Global Compact, amongst others). The six domains include environment, community involvement, business behaviour, human rights, governance, and human resources. In order to determine the relevant weighting, the agency applies three factors in bunches: the nature of stakeholders' rights, interests, and expectations; the vulnerability of stakeholders by sector; and the risk categories for the company: human capital cohesion, operational and organisational efficiency, reputation, legal security, market security, and transparency.

Standard Ethics

As mentioned earlier, the list derived from the SustainAbility list excludes Standard Ethics. The company has its origins in an asset management company that was established in 2001. In 2002 it began issuing sustainability ratings and in 2004 was spun off into a separate entity, losing the asset management arm. This new entity was labelled AEI Standard Ethics EEIG, but in 2014 Standard Ethics Ltd. absorbed the old company and became the first independent sustainability rating agency 'with a standard methodology and a proprietary algorithm' that utilised the issuer-pays model that is prevalent in the credit rating industry. In order to combat the many conflicts of interest that are associated with that remuneration model, the company declares that it does not provide anybody with additional or 'ancillary' services.

Its methodology is determined by its remuneration model. It does not request information via a questionnaire, as its ratings are solicited, so it adopts an 'analyst-driven' model. It does offer unsolicited ratings only 'where it intends to offer stakeholders indices-benchmarks'. Its ratings range from E- to EEE, which aim to represent advancement in both effort and commitment, and from CSR and Ethics through to 'sustainability'. The agency declares that in following the sentiment from the Brundtland Report, it does not use weights or key performance indicators (KPIs). Rather, it uses a unique algorithm. The algorithm is related to a number of elements, including competition (which positively regards how the company competes and negatively regards risky elements like antitrust issues); the position of minority shareholders; managerial scope and policies; and then finally on ESG factors, such as whether an entity's policies are aligned to the Paris COP21 agreement, for example. Whilst it does not overtly use weights in their ratings, it appears that weightings are built into their systems; it states that its basic weights are governance at 33%, stakeholders at 18%, transparency at 14%, market and position at 13%, and environment at 8%, leaving 'other ESG' at 14%.

Now we more know about the leading 'agencies', it is worth just focusing more on the concept of 'agency' in this context. It is likely unhelpful that the term 'sustainable rating agency' is used so much because, in reality, a number of the entities noted earlier are not agencies, but divisions within larger businesses. From this, arguably only Standard Ethics, Vigeo-Eiris, EcoVadis, and Sustainalytics are truly 'agencies' (with Sustainalytics's sale to Morningstar and Vigeo-Eiris's sale to Moody's perhaps removing them out of the equation). However, could Refinitiv be considered an agency in its new form? Can Vigeo-Eiris be considered an agency if it is now under the umbrella of Moody's, with the same question applying to Sustainalytics as under Morningstar? How impactful is the blurring of lines between the credit rating and sustainability rating industries? Perhaps these questions are simply a natural consequence of such a growing and evolving industry. Perhaps, as was noted at the beginning of this section, it is the case that 'corporate sustainability system' (CSS) should be used more to classify this sector, and I will use this terminology moving forward. However, if that is to be the way forward, it must be done in a holistic and updated manner so as to represent the industry properly. Hopefully, this section has added some clarity to this issue.

The future for the industry: misalignment with the true needs of the market

Whilst the previous section attempted to provide some clarification in terms of understanding the marketplace, the litany of issues that surround the industry needs to be addressed as well. There are perhaps two particular issues that will dictate the future trajectory of the sustainability rating industry, and we shall cover them in turn. The first is the methodological divergence that is being witnessed in the industry. The second, which is inherently connected, is the structure of the industry and the impact of increased competition. When these two issues are taken into consideration, it quickly becomes evident that, as an industry, a massive threat is surely not far away from impacting the sector.

On the issue of methodology, and the issues surrounding it within this sector specifically, the literature is emphatic. Escrig-Olmedo et al. argue that the current mechanisms for rating sustainability are inadequate, advancing the notion that there are a number of issues to be resolved (2019, p. 921). First, there is a lack of transparency in that the CSSs do not offer complete information publicly regarding the methodological processes they employ. This has the effect of making comparisons between them difficult. The description of the methodologies in the last section was based on small amounts of investigative research from two legal practitioners via Harvard Business School, small sections of the literature, and a review of the CSSs' websites, which was hardly illuminating except for one or two, like EcoVadis. Second, commensurability is a massive issue. This means that the CSSs may measure the same concept in different ways, depending upon their cultural origins, attitudes to processes, or understanding of materiality. If this is the case, then as the scholars note, if the assessments are not consistent 'the hypothesised benefits of CSR cannot occur'. Another issue to note is that, often, the CSSs do not address the different stakeholders' expectations during their evaluative processes, which the scholars suggest impacts upon their usefulness. It was interesting to note from the SustainAbility reported cited earlier that the CDP offering was considered the most useful by surveyed market participants, despite the CDP's ratings not being independently verified like many of the others. Maybe the CDP have simplified their differentiation so much that it is useful for market participants?

On a positive note, the scholars discuss how the methodologies across the board are apparently evolving to reflect modern standards. For example, focus on the 'E' element of ESG has slowly transformed to consider different standards of what environmental protection looks like, with the Paris COP21 agreement featuring heavily (ibid 925). For the 'S', labour management, human rights, health and safety, and working conditions now feature more heavily than they did ten years ago, whilst for the 'G' element, apart from the base standards of board structure and internal governance mechanisms, elements such as prevention of corruption and bribery now figure much more prominently than they did ten years ago. Yet the focus is predominantly on the divergence and the impact that may have. Chatterji et al. argue that low commensurability means that the CSSs' ratings have a low level of validity and should be treated with caution (2016, p. 1607). Berg et al. produced research recently that goes further, arguing that there exists something they label as the 'rater effect'. This means that for particular CSSs, once a firm performs well in a given category, they are then more likely to perform well in other categories according to that CSS. They also find that, alternatively, if a firm performs poorly in a given category, then that rating is replicated across the board (2019, p. 4). Whilst it could of course be argued that a poorly performing firm would perform, in relation to ESG application, poorly across the board, the level of likelihood suggests that it is the raters' sentiment that determines this, not necessarily the firm's performance. Further still, on divergence across the CSSs, the researchers found a low level of agreement on even standardised elements, such as membership of the Global Compact and CEO/chairperson separation, with the sample of six CSSs showing a remarkable correlation of 0.86 and 0.56 on this topics, respectively (ibid 19). At certain points, the researchers find that there is even negative correlation, meaning that for certain elements – particularly for 'S' elements which tend to be the most subjective – the CSSs are not just diverging, but disagreeing.

The researchers noted earlier spoke to a number of the CSSs they were focusing upon, and the former RobecoSAM stated that one of the reasons for this divergence and also for the 'rater effect' is that sometimes firms may not fully disclose all of the information requested – they may not answer all of the questions on a questionnaire or may only choose to validate certain elements. As not everybody uses issuer-pays as a system, their connection to rated entities is dependent upon the entity, not the rater. Drempetic et al. agree that disclosure lies at the heart of CSS accuracy (2019, p. 16), whilst Hill notes a number of issues relating to disclosure, including the fact that some must disclose because they are regulatory bound to do so, whilst others are not which will have a naturally destabilising effect upon the final ratings (Hill 2020, p. 175). He also raises the point, as I did earlier via introducing the CSSs, that the verification processes for each CSS are very different or sometimes non-existent. Combine this with Hill's understanding that less than 5% of the world's public companies report their emissions rates, and less so on other issues, then we can see that CSSs are a slave to the disclosure rates of the rated entities. Eccles et al. argue that the act of rating is a 'two-way street', in that whilst the character of the rater will impact how they rate, the act of rating then impacts upon those who are rated and affects how they operate, which they see as a positive externality; the mere awareness of being observed (rated) can change one's behaviour, or the so-called 'Hawthorne Effect' (Eccles et al. 2019, p. 5). That may well be the case, but with claims of low validity, the need for more reliability in the data (Țîrcă et al. 2019, p. 10), and companies complaining of a 'tiresome and burdensome' process in meeting all of these different demands (Mohinof and Rogers 2017), the scope for this positive externality seems to be limited, and potentially reducing the more the industry expands.

This leads us to the final issue, which is the structure of the industry itself. The research produced by Berg et al. cited earlier was picked up by the financial press recently (Nauman 2020), with the headline sentiment being that the sustainability rating industry was not suited to the needs of the marketplace. The sentiment is that investors need as much certainty, and as much simplification, as possible. In the credit rating industry, the oligopolistic structure of the industry – for better or worse – provides this, as there is a limited amount of rarely diverging information being pumped into the investment system. In the sustainability universe, this is not the case. A vast number of CSSs are producing information, often on the same subjects, but which are different from each other. The literature is suggesting, ever so slightly, that this makes the industry inefficient. However, if we take this further, it can be suggested that this element not only makes them inefficient but is the element which may see the industry change beyond all recognition. The CSSs provide two elements, mainly, one which the much larger and much more resourced credit rating industry need and one which they could take advantage of if it meant satisfying the first need. The credit rating industry, especially the big two of S&P and Moody's, are making *massive* efforts to be the face of the mainstreaming of ESG into the investment dynamic, via the required third-party verification, or signalling role, that the mainstreaming must contain. Its M&A strategies confirm this, and we saw this earlier when learning about the CSSs. In taking over more of the industry, they can arguably offer the marketplace what it needs – ESG data emanating from fewer sources. The other thing that the CSS universe does is provide information for a universe of indices that provide crucial information for investors; there is no reason that credit rating agencies would not either take this role on board via the purchasing of CSSs, or spin those elements off and create a small and very specialised marketplace for ESG information *for indices*, rather than for *rating* per se. Either way, the threat to the CSS universe as it stands is real and well known.

Conclusion

The drive to incorporate sustainability in the financial psyche since the financial crisis has been a tremendously impactful movement. It has achieved a massive amount in such a small amount of time. However, at some point, the reality of the dynamics required to actually implement that movement had to come to the fore, and it appears that this may be happening shortly. The inherent mechanisms within the concept of investing within a market economy have existed for a long time, and now the 'mainstreaming' of the sustainable movement has moved forward, that inherent dynamic is now impacting upon the development of the movement. Key to that dynamic is the 'signalling' role that is inhabited by 'rating' entities. It is telling that before the move began to accelerate, smaller CSSs were able to slowly develop. Now the acceleration has started to take hold, they are now being forced to adapt and evolve, which brings them into the realm of much larger entities. This is because the 'mainstream' have certain requirements that the luxury of being a niche service provider simply do not fill. The need for measurable and comparable information is key to the mainstream investing universe, and that need will be met before that need is changed. The methodological concerns that the literature has raised feed into a larger required research topic of the requirements of the mainstream vs the dynamics of the marketplace.

References

Aaron Chatterji, Rodolphe Durand, David I Levine, and Samuel Touboul, 'Do Ratings of Firm Converge? Implications for Managers, Investors and Strategy Researchers' (2016) 37 *Strategic Management Journal* 8 1597–1614.

Alexis H Bateman, Edgar E Blanco, and Yossi Sheffi, 'Disclosing and Reporting Environmental Sustainability of Supply Chains' in Yann Bouchery, Charles J Corbett, Jan C Fransoo, and Tarkan Tan, *Sustainable Supply Chains: A Research-Based Textbook on Operations and Strategy* (Cham: Springer 2016).

Ali Shahzad, Nicholas Bartoski, Brandi K McManus, and Mark P Sharfman, 'A Researcher's Guide to Business and Society Archival Datasets' in Abagail McWilliams, Deborah E Rupp, Donald S Siegel, Gunter K Stahl, and David A Waldman, *The Oxford Handbook of Corporate Social Responsibility* (Oxford: Oxford University Press 2019).

Betty M Huber, and Michael Comstock, 'ESG Reports and Ratings: What They Are, Why They Matter' (2017) *Harvard Law School Forum on Corporate Governance* (July 27).

Billy Nauman, 'Heavy Flows into ESG Funds Raise Questions Over Ratings' (2020) *Financial Times* (March 4).

Borja Diez-Cañamero, Tania Bishara, Jose Ramon Otegi-Olaso, Rikardo Minguez, and José María Fernández, 'Measurement of Corporate Social Responsibility: A Review of Corporate Sustainability Indexes, Rankings and Ratings' (2020) 12 *Sustainability* 2153.

Cary Krosinsky, 'Sustainable Equity Investing: The Market-Beating Strategy' in Cary Krosinsky, *Sustainable Investing: The Art of Long-Term Performance* (London: Earthscan 2012).

Cary Krosinsky, and Nick Robins, *Sustainable Investing: The Art of Long-Term Performance* (London: Earthscan 2008).

Christoph Butz, and Jean Laville, 'Ethos Discussion Paper No. 2: Socially Responsible Investment: Avoiding the Financial Materiality Trap' (2007) Discussion Papers, Ethos Foundation.

Christophe Revelli, 'Socially Responsible Investing (SRI): From Mainstream to Margin?' (2017) 39 *Research in International Business and Finance* 711–717.

Daniel Cash, *Regulation and the Credit Rating Agencies: Restraining Ancillary Services* (Abingdon: Routledge 2018).

David A Lubin, and Daniel C Esty, 'The Sustainability Imperative' in Cary Krosinsky, Nick Robins, and Stephen Viederman, *Evolutions in Sustainable Investing: Strategies, Funds and Thought Leadership* (Oxford: John Wiley & Sons 2011).

Diana-Michaela Țîrcă, Anişoara-Niculina Apetri, and Mirela I Aceleanu, 'Sustainability in Finance and Economics' in Magdalena Ziolo, and Bruno S Sergi, *Financing Sustainable Development: Key Challenges and Prospects* (Cham: Springer 2019).

Elena Escrig-Olmedo, Maria A Fernandex-Izquierdo, Idoya Ferrero-Ferrero, Juana M Rivera-Lirio, and Maria J Munoz-Torres, 'Rating the Raters: Evaluating How ESG Rating Agencies Integrate Sustainability Principles' (2019) 11 *Sustainability* 915.

Florian Berg, Julian F Koelbel, and Roberto Rigobon, 'Aggregate Confusion: The Divergence of ESG Ratings' (2019) MIT Sloan School Working Paper 5822–19.

Gilbert S Hedstrom, *Sustainability: What It Is and How to Measure It* (Berlin: Walter de Gruyter 2018).

John Cullis, Philip Jones, and Alan Lewis, 'Ethical Investing: Where Are We Now?' In Morris Altman, *Handbook of Contemporary Behavioural Economics: Foundations and Developments* (Abingdon: Routledge 2015).

John Hill, *Environmental, Social, and Governance (ESG) Investing: A Balanced Review of Theoretical Backgrounds and Practical Implications* (London: Elsevier 2020).

Kate Miles, *The Origins of International Investment Law: Empire, Environment and the Safeguarding of Capital* (Cambridge: Cambridge University Press 2013).

Marguerite Mendell, and Erica Barbosa, 'Impact Investing: A Preliminary Analysis of Emergent Primary and Secondary Exchange Platforms' (2012) 3 *Journal of Sustainable Finance & Investment* 2 111–123.

Michael S Pagano, Graham Sinclair, and Tina Yang, 'Understanding ESG Ratings and ESG Indexes' in Sabri Boubaker, Douglas Cumming, and Duc K Nguyen, *Research Handbook of Finance and Sustainability* (Cheltenham: Edward Elgar 2018).

Nick Robins, 'The Emergence of Sustainable Investing' in Cary Krosinsky, *Sustainable Investing: The Art of Long-Term Performance* (London: Earthscan 2012).

Olaf Weber, 'Finance and Sustainability' in Harald Heinrichs, Pim Martens, Gerd Michelsen, and Arnim Wiek, *Sustainability Science: An Introduction* (Cham: Springer 2015).

Pieter J Trinks, and Bert Scholtens, 'The Opportunity Cost of Negative Screening in Socially Responsible Investing' (2017) 140 *Journal of Business Ethics* 193–208.

Robert G Eccles, Linda-Eling Lee, and Judith C Stroehle, 'The Social Origins of ESG: An Analysis of Innovest and KLD' (2019) *Organization & Environment* 1–22.

Rosella Caré, *Sustainable Banking: Issues and Challenges* (Cham: Springer 2018).

Samuel Drempetic, Christian Klein, and Bernhard Zwergel, 'The Influence of Firm Size on the ESG Score: Corporate Sustainability Ratings Under Review' (2019) *Journal of Business Ethics* 1.

Sanjai Bhagat, *Financial Crises, Corporate Governance, and Bank Capital* (Cambridge: Cambridge University Press 2017).

Scott R Herriott, *Metrics for Sustainable Business: Measures and Standards for the Assessment of Organisations* (Abingdon: Routledge 2016).

Sirkku Hellsten, and Chris Mallin, 'Are "Ethical" or "Socially Responsible" Investment Socially Responsible?' (2006) 66 *Journal of Business Ethics* 393–406.

SustainAbility, *Rate the Raters 2019: Expert Views on ESG Ratings* (New York City: SustainAbility 2019).

Tessa Hebb, 'Introduction – The Next Generation of Responsible Investing' in Tessa Hebb, *The Next Generation of Responsible Investing* (Cham: Springer 2011).

Thomas Johnsen, Mickey Howard, and Joe Miemczyk, *Purchasing and Supply Chain Management: A Sustainability Perspective* (Abingdon: Routledge 2014).

Thomas Walker, Stéfanie D Kibsey, and Stephanie Lee, 'Impact Investing' in Cary Krosinsky, and Sophie Purdom, *Sustainable Investing: Revolutions in Theory and Practice* (Abingdon: Taylor & Francis 2016).

Tim Mohinof, and Jean Rogers, 'SASB and GRI Pen Joint Op-Ed on Sustainability Reporting Synchronicity' (2017) www.sasb.org/blog/blog-sasb-gri-pen-joint-op-ed-sustainability-reporting-sychronicity/.

World Commission on Environment and Development, *Our Common Future* (Oxford: Oxford University Press 1990).

Zabihollah Rezaee, *Business Sustainability, Corporate Governance, and Organisational Ethics* (Oxford: John Wiley & Sons 2019).

Chapter 10

Sustainable public finance system

Daniele Schilirò

Introduction

Sustainable development means that an unlimited number of stakeholders will benefit from positive external results in the economy, society, and environment.

The 2030 Agenda's 17 Sustainable Development Goals (SDGs), as well as the Paris Climate Agreement as emphasized in the United Nations secretary-general's report (UNSG 2019), provides a pathway for a more prosperous, equitable, and sustainable future. In particular, the 2030 Agenda[1] defined 17 SDGs and 169 targets to end poverty, fight inequalities, tackle climate change, and ensure that no one is left behind. In this context, state and public finances can play an important role in the actions to support the implementation of SDGs, although many countries are experiencing a period of strong debt growth.

Nevertheless, governments are striving to fulfill such SDGs. Several countries have implemented policies and national mechanisms to achieve the SDGs contained in the 2030 Agenda. However, there are many difficulties in putting these policies into practice and creating conditions to finance these policies even if sustainable development is established among the principles that govern intervention policies of the European Union (EU) and other regions in the world.

The governance of a public finance system is crucial to ensuring the inclusion of social and environmental aspects. Environmental, social, and governance (ESG) elements should be integral to sustainable finance. Therefore, the issue of financing sustainable development and, in particular, pursuing and achieving a sustainable public finance system, has become a key topic.

This chapter discusses the role of state and public finances in taking actions to support the implementation of SDGs. It stresses the need to design and implement a coherent public finance system for sustainable development, trying to analyze the complex problems and possible solutions to reach this aim. Moreover, attention will be given to public–private partnerships in implementing SDGs. This chapter will also analyze methods to enhance sustainable financing strategies and investments. Public expenditure and taxation are both considered crucial policy tools for environmental protection. The role of taxes in reducing greenhouse gas emissions and removing inefficient fossil fuel subsidies are presented as instruments for influencing and shaping the attitude of companies and households for sustainable development.

Public finance for sustainable development

Sustainable development needs financing. Thus, a sustainable public finance system and finance strategies must be implemented to achieve sustainable development. Currently, progress toward the fulfillment of the 2030 Agenda remains slow and uneven with the risk of missing several SDGs. In fact, millions of people continue to live in extreme poverty. Income inequality within and among countries is rising. At the same time, unsustainable consumption and production patterns are challenging life on our planet (Schilirò 2019). Ripple et al. (2020) highlighted the urgency and gravity that climate change has on our planet, suggesting drastic recommendations that must translate into political and economic decisions. Energy-sector strategies include (1) replacement of fossil fuels; (2) reduction of climate pollutants like fine dust; (3) ecosystem restoration; (4) change in food consumption; (5) conversion toward a carbon-free economy (the climate crisis will have a strong negative impact on finance); and (6) sustainable demographic policy.

Individual states and international cooperation between states are decisive in reaching the SDGs. Public finance plays a key role in taking actions to support and implement the achievement of SDGs (e.g., reducing poverty, establishing good health and well-being, providing a quality education, tackling climate change, etc.). In particular, public expenditure is crucial in financing investments and technologies conducive to environmental protection. It also stimulates private investments through, for instance, the cofinancing of projects with the private sector. At the same time, tax policy can play a decisive role in reducing greenhouse gas emissions and removing inefficient fossil fuel subsidies. However, the achievement of sustainable development needs a great amount of financial expenditures that public finances, although important and necessary, are unable to cope with, as well as due to the high and growing public debt that currently characterizes many economies (developed, developing, and emerging).

In the private finance domain, sustainable funds have been growing fast. Sustainable finance refers to any form of financial service integrating ESG criteria into the business or investment decisions for the lasting benefit of both clients and society at large. However, the investing industry focuses primarily on the environmental area, but not so much on the social area. Private investors are becoming aware of the challenges related to climate change. Therefore, climate risks must be considered when assessing profit expectations. For this purpose, investors are developing ways to measure climate risks in a more reliable manner by better redirecting capital flows into sustainable projects so that the financial sector appears committed in the fight against climate change. Thus, sustainable finance is an essential tool to accelerate climate action and effectively manage the risks and opportunities associated with climate change.

Although private finance is called upon to take a leading role in supporting the fight against climate change through sustainable funds, this may not be enough if maximizing profits and generating growth continues to be the main premise. Thus, a change of the economic paradigm from a neoclassical approach is needed. Behavioral economics, in particular, can suggest insights to public policymakers regarding how the environment and the context can provide conditions for better choices.

Behavioral economics can reveal aspects of developed societies that are unsustainable and contribute to fulfilling the SDGs. In fact, contributions of behavioral economics can redefine the goals of the economic agents (i.e., enterprises, financial institutions, consumers, and governments and public authorities) from behaviors aimed at maximizing target variables (i.e., profits, levels of consumption, rates of growth) to alternative behaviors compatible with the SDGs and requirements of the Paris Agreement[2]. An example of a positive move in this direction is when Israel decided to develop and update indicators of well-being, sustainability, and national resilience in domains like quality of employment, skills and higher education, environment, social well-being, community, and culture (United Nations 2019).

Therefore, there is a prior condition concerning the definition of target variables in public and private finance which must be aligned with SDGs. Accepting a new approach based on a long-term vision to preserve the natural capital and the environment will allow us to explore sustainable finance aimed at achieving the SDGs and the goals of the Paris Agreement.

Public and private finance for sustainable development

According to UNEP (2018), financing the SDGs and the Paris Agreement requires investments amounting to trillions of dollars per year for the next decade and beyond. Therefore, the SDG targets cannot be reached without private-sector involvement.

The economic literature (e.g. Gardiner et al. 2015; Cheng et al. 2020; Horrocks et al. 2020) agrees that a public-sector partnership with the private sector is a necessary, although not a sufficient, condition to implement sustainable development and its objectives. Though the public finance system must have a role in pursuing and incentivizing the SDGs, public-sector entities are not able to meet the sustainable development objectives on their own; thus, much of the finance must come from private sources due to the scarcity of public finance. Particularly, developing and emerging market countries have relatively small public sectors and tax revenue bases and limited fiscal space, as Djankov and Panizza (2020) point out.

Yet financing the SDGs is a complicated task. It needs unprecedented coordination between public and private sectors. A great challenge is to align the huge amount of money invested daily in capital markets with the SDGs. All this also requires significant efforts to implement reforms of global financial regulations and financial institutions, as well as strong commitment from all stakeholders.

A move in this direction seems to be the Action Plan on Sustainable Finance by the European Commission (European Commission 2018a). This plan makes sustainability a part of financial decision-making. The plan has three broad aims. First, it aims to reorient capital flows toward a more sustainable economy. In particular, the EU intends to involve more private capital for environmentally sustainable investment.[3] Second, it aims to maintain sustainability in risk management through the integration of its Action Plan on ESG factors within the asset management. Third, it aims to foster transparency and sustainability. In addition, the European Commission (2018b) issued a proposal to facilitate sustainable investment and set the foundation for an EU framework that puts the ESG considerations at the heart of the financial system. This proposal establishes the conditions and framework to gradually create a unified classification system (or taxonomy) on what can be considered environmentally sustainable for investment purposes.

Also, the European Commission, in the Annual Sustainable Growth Strategy 2020 (European Commission 2019b: 1), stated that "Economic growth is not an end in itself. An economy must work for the people and the planet". Furthermore, this growth strategy consists of the European Green Deal, placing sustainability and the well-being of citizens at the center of any action (European Commission 2019a).[4] This could represent a change of pace, especially because the European Commission aims to mobilize investment of €1 trillion over 10 years through public and private money to finance its flagship project, the European Green Deal.

On January 14, 2020, the European Commission presented the Sustainable Europe Investment Plan to shift the European economy to net-zero CO_2 emissions by 2050. It aims to protect coal-dependent regions from absorbing the brunt of changes aimed at fighting climate change. All European regions will require funding. The European Green Deal includes the Just Transition Mechanism (European Commission 2020a), a key tool to ensure that the transition toward a climate-neutral economy happens in a fair way, leaving no one behind. The mechanism provides targeted support to mobilize at least €100 billion (2021–2027) in the most affected regions to alleviate the socioeconomic impact of the transition. The plan of the European

Commission president also aims to introduce a climate law to ensure that all European policies are geared toward the climate change neutrality objective. The target is an agreement between member states to cut emissions by 40% between 2017 and 2030.

Criticisms of this ambitious project have not been lacking. It has been pointed out that a grand plan for a distant future raises skepticism, especially for leaders who face re-election every four or five years. A 2050 target is certainly not binding. Even more, opposition by fossil fuel–producing member states, energy-intensive sectors, trade-sensitive industries, and car-dependent households will be fierce (Pisani Ferry 2019). An issue concerns the realistic aspect of the plan. More specifically, the EU cannot rule directly on the member states' energy mix, housing standards, taxes, and public investment. Therefore, designing and enforcing a common EU strategy will be not an easy task. Moreover, tough regulations and dissuasive taxation are difficult weapons to implement. A realistic tool to fight climate change becomes green finance. However, the latter may be insufficient to cope with the problems of decarbonization. Therefore, the transition to carbon requires lifestyle changes.

Another type of criticism has been brought forward by Huber (2020). She points out that the perspective of the plan is that humans can make nature work for them through technology and transform climate change mitigation and containment into jobs and growth, rather than radically reimagining our way of life in synchronization with the nature of which we are part.

Despite these criticism and problems, the European Green Deal should represent, according to the European Commission, the EU's new defining mission to lead the EU toward a path of sustainability.

In addition to the EU, development finance institutions around the world are increasingly using their balance sheets to leverage private capital alongside measures to de-risk investments by encouraging ranging policy and institutional developments. There is relevance regarding meritorious initiatives from institutions at international levels. Yet the mobilization of both public and private resources and investments falls short.

Furthermore, the role of private finance is controversial. Many cases, both in emerging and developing economies, are not aligned with SDGs. In addition, they are not compatible with climate-resilient developments. This represents a major obstacle to the advancement of the SDGs.

A possible solution for the cooperation between public and private finance is the development of innovative financing. This refers to financial solutions to development challenges that remain insufficiently addressed by traditional aid flows. In this view, there are two subcategories of innovative financing (Benn and Mirabile 2014). First, some innovative sources generate new financial flows for sustainable development that may come from various economic sectors. Second, some innovative mechanisms maximize the efficiency, impact, and leverage of existing resources. Particularly, innovative financing mechanisms can blend in different ways within public finance. This can offset risks and subsidize or incentivize private lending and investment.

Such financial innovations are vital, serving as the subject of experimentation and growing practice. However, limits to the volume of public finance that can be redirected to this purpose represent a major constraint to the rapid scaling of blended financing. Certainly, reforms in the real economy can integrate these financing mechanisms as policy, market, and technological developments change the relative prices, risks, and returns to sustainability-aligned financing. One example is giving subsidies to private capital. This provides finance for investments like the deployment of renewable energy in which improved returns to private capital are secured through direct public subsidies. Another example is the imposition of surcharges on electricity consumer prices. Although some of these changes are visible, such as the falling cost of clean energy systems, the scale of redeployment of private capital often remains inadequate.

A shift toward sustainable finance and the greening of the financial system

Both the 2015 Paris Climate Agreement and the SDGs require a new generation of innovation from the financial system. There is a growing awareness and increased actions by financial institutions across the banking and insurance sectors, capital markets, and institutional investment. The issue is to overcome barriers like mispricing, short-termism, and low levels of awareness and capability that prevent the scaling up of good practices. Harnessing the financial system will be essential for achieving a successful transition to a low-carbon, inclusive, and sustainable model of development.

Sustainable finance involves the integration of ESG factors across the financial system. Its goal is to strengthen resilience, target capital allocation, and improve accountability. The focus is mainly on the environmental dimension of financing sustainable development, termed "green finance". Green finance aims to guarantee the finance for needed environmental projects and make more sustainable finance (or "greener finance").

There is growing recognition that ESG factors are vital for value creation. Environmental threats like climate change and water scarcity are creating risks to financial assets and challenges, particularly for the insurance sector.[5] Banks, capital markets, and institutional investors are incorporating environmental and social factors into capital allocation decisions. Public finance will be key to enabling this shift. Yet the bulk of the capital required will need to come from the private sector. At the same time, international policy cooperation is occurring. Increasingly, finance ministries, central banks, and regulators are identifying how sustainability factors impact financial stability and long-term investment.

However, green finance has not yet achieved a systematic impact across the financial mainstream as a result of a number of challenges. Particularly, limited access to finance for enterprises, especially for small and medium enterprises, is constraining their participation in the green economy. Moreover, unpriced environmental externalities can tilt the risk/return profile away from sustainable finance. Financial culture places an insufficient emphasis on the skills and capabilities required to respond to the conditions of sustainable development in many countries. Last, financial decision-making does not adequately consider long-term challenges like climate change. Together these factors have led to insufficient flows of capital to the green economy, leaving the sustainable development and climate goals unrealized.

Over the past decade, there have been increasing efforts by financial institutions to align the financial system with long-term sustainable development. This is being driven by the growing acknowledgement of the value of sustainability factors for efficient capital allocation to the real economy, the delivery of risk-adjusted returns, the management of emerging threats, and the strengthening of economic governance.

Regarding a sustainable financial system, in addition to the SDGs and the Paris Agreement, there is the "action agenda" of the Financing for Development Conference (United Nations 2015) held in Addis Ababa. This agenda focused on steps to increase domestic and international mobilization for developing countries in terms of public and private capital. One of its conclusions was to "strengthen regulatory frameworks to better align private sector incentives with public goals, including incentivizing the private sector to adopt sustainable practices, and foster long-term quality investment" (United Nations 2015: 17) from both domestic and international institutions. Regarding investments and infrastructure financing, the agenda specifies that "both public and private investment have key roles to play, including through development banks, development finance institutions and tools and mechanisms such as public-private partnerships and blended finance" (United Nations 2015: 24).

To deploy capital at the scale and speed required, a number of interlocking elements have to be in place. First, policy action in the real economy is needed to remove market failures like unpriced pollution and resources. As a matter of fact, although progress has been made on internalizing externalities into market prices, serious market failures remain worldwide. Risk emerges without effective pricing of scarce natural capital. Adjusted returns for sustainable finance are likely to be inadequate to attract sufficient capital.

Second, the effective deployment of public finance that provides public goods and stimulates private action is also needed. Public finance is essential to deliver collective goods that the market cannot provide. It also stimulates private action through incentives and subsidies. However, public finance remains insufficient in all developing countries.

Third, action is needed within the financial system to remove market and institutional barriers that can prevent the efficient allocation of capital to sustainable development. These include misaligned incentives, short-termism, inadequate risk management, insufficient transparency, and poor stewardship.

Among the actions to be undertaken, there is the need of additional global clean energy investments to extend the time horizon concerning infrastructure. At the same time, financial markets and policy can suffer from a "tragedy of horizon", discounting future risks in today's decisions and risking irreversible damage.

The role of public finance for sustainability

Public finance is critical for the delivery of public goods that the market will not deliver. This can take the form of fiscal expenditure and subsidies. However, public finance is scarce. Thus, it needs to be used with a focus on efficiency and effectiveness. Proposed expenditures require the identification of a sustainable source of funding. Moreover, public finance usually has a longer time horizon and different risk/return expectations with respect to the private sector.

Public financial institutions have also contributed to coinvest and mitigate risks. They have often been at the forefront of innovation in sustainable finance, particularly in terms of implementing risk management frameworks, financing environmental assets (such as through the issuance of green bonds), and developing frameworks for reporting and disclosure.

The United Nations Development Programme (UNDP 2015), through the initiative of Governance of Climate Change Finance (GCCF), has contributed to the development of tools and methodologies to design a sustainable public finance system and help country decision makers improve their spending on climate change. More recently, with the *guiding note*, UNDP (2019) aims to guide governments and institutions through the process of creating a Climate Change Financing Framework (CCFF).

Regarding the design of a sustainable financial system, it is interesting to look at the Italian case. UNEP (2016), in particular, identified 18 options across the Italian financial system. The report stressed the importance of financing sustainable infrastructure and energy efficiency. Also, it required to review tax expenditures in order to remove progressively environmentally harmful subsidies in the energy sector. As a matter of fact, Italy, as documented by ENEA (2018) and the European Commission's Country Report on Italy (European Commission 2020b), is on track to achieve its 2020 climate and energy targets; moreover, the energy efficiency policies adopted aim to achieve 30% energy savings by 2030 compared to the expected consumption at that date (ENEA 2018). However, the country must considerably review environmental tax expenditures and, in general, fiscal policy. As far as sustainable infrastructure is concerned, Italy still lags behind with respect to most of the EU countries. There is therefore a clear delay in the country's ability to respond to the new sustainable infrastructure needs and to its financing.

Furthermore, despite the uncertainty and disagreement on the economic costs of climate change and the impact of mitigation measures on economic growth, there has been an increasing acceptance of the argument that economic growth based on unplanned exploitation of natural resources is not sustainable (Schilirò 2019). A public finance that favors the transition to a low-carbon economy can have a positive effect in terms of sustainable development, even if it does not avert the negative impact of climate change by itself. At the same time, fiscal policy action for climate mitigation aims to use environmental tax reform to align energy prices to fully reflect climate externalities.

Developing countries, in particular, need immediate action to protect their environment, move toward a low-carbon economy, and manage their natural capitals to ensure sustainable development and lasting economic growth. A possible solution is, as mentioned, the greening of the public financial system. Managing a greening public financial system means embedding both environmental conservation and sustainable management of natural capital as a decision-making and evaluation criterion.

Tax policy to change consumption and production patterns

Human demands on natural resources have outpaced what can be produced. A shift toward a more sustainable development path is dependent on changes in current patterns of both consumption and production. Ripple et al. (2020) underlined the need for a change in those patterns if we want to preserve our planet from the risks of climate change, realize the SDG targets, and look at a sustainable future. In this complex and challenging scenario, a sustainable public finance system that is dependent on tax policy and public expenditure management can make an important contribution. In fact, it can reinforce a country's infrastructure, reduce poverty and inequality, and contribute to safeguarding the environment and natural resources. Tax policy, in particular, is intrinsically linked to sustainable development, as taxation provides the revenue that states need to mobilize resources. The role of taxation is crucial to achieve the SDGs. Therefore, it should be designed to reduce inequality and promote inclusive growth. The SDGs, in fact, require significant investment. Countries must boost the effectiveness of their tax system to generate the necessary resources.

The empirical literature suggests that tax policy should be flexible and capable of adjusting to socioeconomic changes as the fiscal environment in which they operate evolves. Therefore, the relationship between tax and sustainable development is complex. It is nearly impossible to find a simple framework and universal solution. A realistic solution is to single out the variety of fiscal tools and polices suited to the context of each country. However, a general rule to apply in order to have a tax system consistent with sustainable public finance is enhancing transparency in all financial transactions between governments and companies to avoid illicit financial flows. However, the real challenge of a tax policy aimed at sustainability is to incentivize the households and the private sector to contribute to better environmental, social, and economic outcomes.

Climate change is mostly driven by energy use and land use changes. Therefore, the challenge for a system of taxation is to adopt measures to influence and modify energy production and consumption. At the same time, it must guarantee food security and water security because these aspects are inextricably linked. Moreover, adopting the appropriate environment fiscal policy includes tools aimed at taxing activities that cause environmental damage and excessive use of natural resources, redirecting public and private investments destined for the exploitation of fossil fuels toward low-carbon alternatives.[6] Taxes that reduce greenhouse gas emissions and remove inefficient fossil fuel subsidies go in this direction. Such fiscal tools are also aimed at fueling the development of the green economy, giving signals to the markets to encourage long-term sustainable investments.

This fiscal policy could increase production costs in the countries that adopt them, in particular by penalizing certain sectors. This is why it is necessary that this fiscal policy be adopted in a coordinated way at an international level to avoid penalization in terms of competitiveness of those countries that adopt stricter environmental policies. In addition, clear, transparent, and consistent international standards of taxation are essential for cross-border trade, business investment, jobs, and sustainable development.

Environmental tax policies play an important role in the transition to a green economy. Among the tools, environmental taxes are not the only instrument. For example, incentives for investments in renewable energies (i.e., solar, geothermal, wind) represent effective solutions. They can guarantee a fixed tariff for all the energy produced. Moreover, public support can gradually be eliminated with increases in the share of renewables in the energy mix, as well as the affirmation of their competitiveness with respect to conventional fuels. Such initial investment support is essential to redirect capital flows otherwise destined for carbon-intensive energies.

It is also true that subsidies can create distortions and unwanted effects. Some forms of subsidy for the production of clean energy can create distortions in the market in the following ways: (1) rewarding a particular technology; (2) directing investments toward it; and (3) slowing down the development and diffusion of other green technologies or prevent them from happening in a more convenient and efficient manner. However, governments must think more creatively, going beyond the exclusive use of environmental taxation, particularly in developing countries.

In combination with the tax reform, transparent communication and commitment to stakeholders and better monitoring need to be put in place. As Weber (2015) emphasized, there is a status quo bias in the attitude toward climate change. This is why it is important to activate initiatives to obtain *behavioral changes* to encourage the adoption of actions with high energy efficiency and low carbon emissions. These *behavioral changes* can be based on a "new conceptualization of human happiness, away from the current model that is consumption-based which puts us on a hedonic treadmill and endangers the global climate and environment" (Weber 2015: 577).

In addition, a growing body of research indicates that consumer choices and behavior are, to a large extent, driven by cognitive biases and heuristics (Gigerenzer and Goldstein 1996; Gigerenzer et al. 1999; Thaler and Sunstein 2008; Tversky and Kahneman 1974). For example, people use mental shortcuts to cut through complexity. They dislike losses more than they like gains (loss aversion) (Kahneman and Tversky 1979), evaluate things in relative rather than in absolute terms, prefer lower-value certainties over higher-value risks (risk aversion) (Laibson 1997; Loewenstein and Prelec 1992), and are influenced by people around them (conformity effect) (Kim and Hommel 2015).

Yet these cognitive biases are often overlooked by policymakers, who seek to use the public finance system to promote energy efficiency and conservation of natural resources. However, it is important to take these psychological aspects into account when developing strategies for motivating pro-environmental behavior in the areas of public finance and, of course, in consumption and production.

Since *behavioral changes* are unavoidable to overcome the absolute emergency of climate change and environmental problems, as well as go in the direction of fulfilling the SDGs, adequate policy actions must be taken. Thus, the setup of a sustainable public finance system and relative public policies must go in that direction.

Conclusions

Public finance for sustainable development has been the key topic of this chapter. In particular, the chapter focused on the role of public finance systems in taking adequate actions to support the implementation of the SDGs, highlighting complex problems and seeking to single out possible solutions. A key issue of the analysis is the necessary public-sector partnership with the private financial sector, considering the limited public resources and given the historically high levels of public debt in developing and emerging economies. The shift toward sustainable finance and greening of the financial system is another necessary step to achieve sustainable development goals. This shift recognizes the importance of ESG factors across the financial system. It also implies a change in the economic paradigm from a neoclassical approach, in particular toward theoretical frameworks like those implied by the behavioral economics approach. This latter allows one to redefine the goals of the economic agents that become compatible with the SDGs and the requirements of the Paris Agreement.

In designing a public financial system, a fundamental issue is how to shift consumption patterns to favor the implementation of the environmental goals and, therefore, a sustainable future. Tax policy and environmental taxation represent important instruments to influence and shape the attitude of households and companies for sustainable development.

The chapter also highlighted the role of taxes in reducing greenhouse gas emissions and removing inefficient fossil fuel subsidies. Yet the chapter stressed that to face the absolute emergency of climate change and environmental deterioration, it is necessary to change human behavior, even if changing human behavior through public policy is challenging and complex.

In conclusion, a sustainable public finance system, as well as the relative tax policy and public expenditure, must go in the direction of fulfilling the SDGs in the 2030 Agenda. Such a sustainable public finance system and the relative fiscal policies should aim to reduce inequality, change consumption patterns, finance investments and technologies conducive to environmental protection, and take measures to mitigate the impact of climate change and promote inclusive growth. These are the open challenges for the future.

Notes

1 The 2030 Agenda consists of four parts: (1) political declaration; (2) set of 17 SDGs and 169 targets; (3) means of implementation; and (4) framework for follow-up and review of the agenda.
2 Simon's notion of satisficing behavior (Simon 1972, 1978) can represent a possible alternative within the behavioral economics approach. See also Schilirò (2018).
3 The European Commission recognized that public money is not enough to meet the European needs of around €180 billion in extra yearly investments over the next decade to meet the targets required in the Paris Agreement.
4 According to the Green Deal, the European Union will have zero economic impact in 2050. It will also aim to decarbonize the energy sector. It will support the industry to innovate and become a world leader in the green economy. It will

introduce forms of private and public transport that are cleaner, cheaper, and healthier. Finally, it will aim to renovate buildings and help people reduce energy bills and the use of energy.

5 We speak of a "green swan" as a phenomenon linked to climate change. It is a possible cause that determines a crisis in the financial markets.

6 This analytical framework is based on a competitive strategy aimed at adopting low-carbon technologies to favor the development of a green economy (Carfì and Schilirò 2012).

References

Benn, J. and Mirabile, M., 2014. Innovating to finance development. In E. Solheim, ed. *Development Co-Operation Report 2014: Mobilising Resources for Sustainable Development*. Paris: OECD, 177–185. https://doi.org/10.1787/dcr-2014-en.

Carfì, D. and Schilirò, D., 2012. A coopetitive model for the green economy. *Economic Modelling*, 29(4), 1215–1219.

Cheng, Z., Wang, H., Xiong, W., Zhu, D. and Cheng, L., 2020. Public – private partnership as a driver of sustainable development: Toward a conceptual framework of sustainability-oriented PPP. *Environment, Development and Sustainability*. https://doi.org/10.1007/s10668-019-00576-1.

Djankov, S. and Panizza, U., 2020. Developing economies after COVID-19: An introduction. In S. Djankov and U. Panizza, eds. *COVID-19 in Developing Economies*. A Vox.org Book. London: CEPR Press, 8–23. Retrieved from file:///C:/Users/User/Downloads/Covid-19_in_developing_economies.pdf.

ENEA, 2018. *Energy Efficiency. Annual Report*. Rome: ENEA – National Agency for New Technologies, Energy and Sustainable Economic Development. Retrieved from www.enea.it/it/seguici/pubblicazioni/pdf-volumi/2018/raee-2018-executive summary-en.pdf.

European Commission, 2018a. *Action Plan: Financing Sustainable Growth*. Brussels, 8.3.2018 COM 97 Final. Brussels: European Commission. Retrieved from https://ec.europa.eu/info/publications/sustainable-finance-resources_en.

European Commission, 2018b. *Proposal for a Regulation of the European Parliament and of the Council on the Establishment of a Framework to Facilitate Sustainable Investment*. Brussels, 24.5.2018, COM (2018) 353 Final. Brussels: European Commission.

European Commission, 2019a. *The European Green Deal*. 11.12.2019, COM (2019) 640. Brussels: European Commission. Retrieved from https://ec.europa.eu/info/sites/info/files/european-green-deal-communication_en.pdf.

European Commission, 2019b. *Annual Sustainable Growth Strategy 2020*. 17.12.2019, COM (2019) 650 Final. Brussels: European Commission. Retrieved from https://ec.europa.eu/info/sites/info/files/2020-european-semester-annual-sustainable-growth-strategy_en.pdf.

European Commission, 2020a. *The Just Transition Mechanism: Making Sure No One Is Left Behind*. Brussels: European Commission. doi:10.2775/19010.

European Commission, 2020b. *Country Report Italy 2020*. Brussels, 26.2.2020 SWD (2020) 511 Final. Brussels: European Commission. Retrieved from https://ec.europa.eu/info/sites/info/files/2020-european_semester_country-report-italy_en.pdf.

Gardiner, A., Bardout, M., Grossi, F. and Dixon-Declève, S., 2015. *Public-Private Finance for Climate Finance*. TemaNord2015:577. Copenhagen: Nordic Council of Ministers.

Gigerenzer, G. and Goldstein, D., 1996. Reasoning the fast and frugal way: Models of bounded rationality. *Psychological Review*, 103(4), 650–669. doi:10.1037/0033-295X.103.4.650.

Gigerenzer, G., Todd, P. M. and ABC Research Group, 1999. *Simple Heuristics That Make Us Smart*. Oxford: Oxford University Press.

Horrocks, P., Boiardi, P. and Bellesi, V., 2020. Shifting public and private finance towards the sustainable development goals. *OECD Development Matters*, 9 January. Retrieved from https://oecd-development-matters.org/2020/01/09/shifting-public-and-private-finance-towards-the-sustainable-development-goals/.

Huber, D., 2020. *The New European Commission's Green Deal and Geopolitical Language: A Critique from a Decentring Perspective*. IAI Papers 20|06, April. Roma: Istituto Affari Internazionali. Retrieved from www.iai.it/sites/default/files/iaip2006.pdf.

Kahneman, D. and Tversky, A., 1979. Prospect theory: An analysis of decision under risk. *Econometrical*, 47(4), 263–291. doi:10.1007/BF0012257.

Kim, D. and Hommel, B., 2015. An event-based account of conformity. *Psychological Science*, 26(4), 484–489. doi.org/10.1177/0956797614568319.

Laibson, D., 1997. Golden eggs and hyperbolic discounting. *Quarterly Journal of Economics*, 112(2), 443–477. doi:10.1162/003355397555253.

Loewenstein, G. and Prelec, D., 1992. Anomalies in intertemporal choice: Evidence and an interpretation. *Quarterly Journal of Economics*, 107(2), 573–597.

Pisani Ferry, J., 2019. *Europe's New Green Identity*. Project Syndicate Commentary, 30.12. 2019. Prague: Project Syndicate. Retrieved from www.project-syndicate.org/commentary/european-union-climate-neutrality-questions-by-jean-pisani-ferry-2019-12.

Ripple, W. J., Wolf, C., Newsome, T., Barnard, P. and Moomaw, W. R., 2020. World scientists' warning of a climate emergency. *BioScience*, 70(1), 8–12. https://doi.org/10.1093/biosci/biz152.

Schilirò, D., 2018. Economic decisions and Simon's notion of bounded rationality. *International Business Research*, 11(7), 64–75.

Schilirò, D., 2019. Sustainability, innovation, and efficiency: A key relationship. In M. Ziolo and B. S. Sergi, eds. *Financing Sustainable Development*. London: Palgrave, 83–102.

Simon, H., 1972. Theories of bounded rationality. In C. B. McGuire and R. Radner, eds. *Decision and Organization. A Volume in Honor of Jacob Marschak*. Amsterdam: North-Holland, 161–176.

Simon, H., 1978. Rationality as a process and a product of thought. *The American Economic Review, Papers and Proceedings*, 68(2), 1–16.

Thaler, R. H. and Sunstein, C. R., 2008. *Nudge: Improving Decisions About Health, Wealth, and Happiness*. New Haven, CT: Yale University Press.

Tversky, A. and Kahneman, D., 1974. Judgment under uncertainty: Heuristics and biases. *Science*, 185(4157), 1124–31. doi:10.1126/science.185.4157.1124.

UNDP, 2015. *Budgeting for Climate Change: How Governments Have Used National Budgets to Articulate a Response to Climate Change, Governance of Climate*

Change Finance. Bangkok: United Nations Development Programme. Retrieved from http://unepinquiry.org/wp-content/uploads/2015/11/The_Financial_System_We_Need_EN.pdf.

UNDP, 2019. *Climate Change. Knowing That You Spend. A Guidance Note for Governments to Track Climate Finance in Their Budgets*. New York: United Nations Development Programme. Retrieved from www.undp.org/content/dam/undp/library/planet/climate-change/RBAP-DG-2019-Climate-Budget-Tagging-Guidance-Note.pdf.

UNEP Inquiry, 2016. *Design of a Sustainable Financial System. Financing the Future – Report of the Italian National Dialogue on Sustainable Finance*. Geneva, Switzerland: UN Environmental Programme. Retrieved from https://unepinquiry.org/.

UNEP Inquiry, 2018. *Making Waves. Aligning the Financial System with Sustainable Development*. Geneva, Switzerland: United Nations Environmental Programme. Retrieved from www.givingcompass.org/wp-content/uploads/2018/05/Making_Waves_lowres.pdf.

United Nations, 2015. *Addis Ababa Action Agenda of the Third International Conference on Financing for Development*. New York: United Nations Department of Economic and Social Affairs Financing for Development Office.

United Nations, 2019. *Implementation of Sustainable Development Goals. National Review Israel 2019*. United Nation Sustainable Development Goals-Knowledge Platform. Retrieved from https://sustainabledevelopment.un.org/content/documents/23576ISRAEL_13191_SDGISRAEL.pdf.

UNSG, 2019. *Roadmap for Financing the 2030 Agenda for Sustainable Development 2019–2021*. New York: United Nations Secretary General's. Retrieved from un.org/sustainabledevelopment/sg-finance-strategy.

Weber, E., 2015. Climate change demands behavioral change: What are the challenges? *Social Research*, 82(3), 560–580.

Chapter 11

Sustainable investing

Ria Sinha and Manipadma Datta

Introduction

Some of the cases of increasing materiality of environmental, social and governance (ESG) factors include the Deepwater Horizon oil spill in which the oil major British Petroleum (BP) recorded a US$53.8 billion pre-tax charge, imposition of €27.4 billion in penalties and fines upon Volkswagen for rigging 11 million diesel vehicles to pass emission tests and the very recent privacy breach of personal data by Facebook in 2018[1]. These are some of the glaring examples where breaches of ESG or sustainable issues have had adverse impacts on the finances of companies. Evidence points to the fact that ignoring sustainability factors can lead to adverse consequences for companies. Instead, the academic literature supports the proposition that organizational entities can leverage sustainability and ESG risks to generate financial returns through phased integration.

There has been a growing consensus of return generating capability of ESG factors. Recognizing the value-generating potential of sustainable investments, firms are increasingly managing ESG risks by applying proper integration techniques and approaches. There is much evidence of increased recognition of sustainable investing and readjusting portfolios by institutional investors (Uzsoki, 2020). Interestingly, sustainability is gaining widespread attention among investors with larger portfolios ($100 billion and above) (Schroders, 2018). Major asset managers such as Amundi and Blackrock have heavily increased their dependence on ESG-themed investments and have aligned their portfolios accordingly. Due to increased evidence of generating higher returns from sustainable investments, pension funds have started to consider ESG issues in processes to avoid breaching their fiduciary duties (Uzsoki, 2020). However, the extent to which ESG issues are integrated in a portfolio depends on the levels of awareness and ESG-related expertise. In a nutshell, the extent of creating a sustainable ESG market depends on the following:

- The level of awareness and acceptance related to ESG;
- Well-grounded research of the impact of ESG factors on firm-level and portfolio performance, i.e. assessment of ESG related risks;

- Availability of well-defined ESG accounting and measuring frameworks;
- Presence of sustainable investment choices in the market;
- Affiliations to ESG-related and responsible investing frameworks such as UN Principles for Responsible Investment (UNPRI), Climate Principles, Equator Principles and the like.

The rest of the chapter will elucidate on the drivers, trends, various evaluation techniques used by investors to conduct research in responsible investing, instruments based on this ideology and the barriers which deter its spread and recommendations.

Evolution of sustainable investing

The increasing prominence of sustainable investing can be gauged from the fact that it has gradually evolved from being solely profit making to be value adding. The process started in early 1500 in the name of ethical investing. Initially the movement was motivated by religious sentiments (Sparkes, 2003). Ethical investing was practiced by religious believers of Judaism, Christianity and Islam, who refrained from investment in certain activities to align with their faith. The religious places of worship, such as churches, played a prominent role in the development of ethical investment products (Benijts, 2010; McCann et al., 2003; Lydenberg, 2002). They specifically advocated against activities, namely gambling, tobacco, alcohol and the like. The process gained recognition with the establishment of the Pioneer investments and the Pioneer Fund in 1928 to enable investors to avoid investments in companies involved in gambling, tobacco and alcohol. Later, 1953 marked the launch of the book on *Social Responsibilities of the Businessman* by Howard R. Bowen, who embarked on issues such as corporate social responsibility for the first time. This seminal book harped on pertinent questions involving responsibilities of the businessmen towards the society and people at large. In 1970, the Securities and Exchange Commission (SEC) began permitting socially responsible issues to appear on proxy ballots after a landmark court case engaging the SEC and the Medical Commission on Human Rights. However, the modern institutionalization of ethical exclusions began with the establishment of the Pax World Fund in 1971, the first socially responsible investment fund in the United States. As this year also witnessed the Vietnam War, Pax offered an alternative investment option to those opposed to the weapons production of nuclear and military arms. In 1977, the movement became intensively globalized through the Sullivan Principles which encouraged divestment and ultimately forced business in South Africa to draft a charter calling for an end to apartheid.

The period from 1960 to the mid-1990s experienced an increase in socially responsible investments, which was mostly values based or an exclusionary investment approach that took account of corporate social, ethical and environmental behavior of the firms. In 1985, the Social Investment Forum was established to advance investment practices that factor in ESG considerations. Later in 1987, with the commissioning of the Brundtland Report, 'Our Common Future', the importance of responsible investing in accomplishing sustainable development was realized.

In the period after the 1960s, socially responsible investing (SRI) emerged as a new concept which had certain differences from ethical investments, i.e. investments driven by certain religious beliefs. The established rationale behind SRI is to make investments which will address societal and community deficits (Camilleri, 2015a; Martí-Ballester, 2015; Nilsson, 2009; Ogrizek, 2002). Although prior to 1990 the focus of SRI was more on labor rights, social accountability of business, women equality rights, rights of the minority communities and other related issues, it was primarily in the late 1990s that SRI started to focus on the sustainable development of the environment (Richardson, 2008; Brundtland, 1989). Up to the mid-1990s, SRI was referred to as a value-based or exclusionary investment approach that primarily took into account the corporate social, ethical and environmental behavior of the firms. Over a period of time as SRI developed more, it shifted away from the emphasis on ethics and more towards incorporation of ESG considerations into investment decisions, thereby transforming itself into an investment strategy which also seeks returns. The evolution of sustainable investing is stated in Table 11.1.

Types of sustainable investing

Several terminologies are interchangeably used to highlight different forms of sustainable investing. However, intricate differences exist between the terms based on sustainability issues considered, the screenings applied, the types of integration across asset classes and ownership practices. Table 11.2 elaborates upon the different types of sustainable investing across the globe.

Relevance of sustainable investing to institutional investors

Relevance of international frameworks

The UNPRI is by far the biggest project undertaken for reporting of sustainability issues by institutional investors. They are a guidance to investors to fulfil their fiduciary duties as well as reap the returns from sustainable investing. These principles promote sustainable investing practices among investors, thereby leading to the formation of sustainable financial markets. Sievänen et al. (2012) finds evidence of achieving higher levels of environmental governance by adhering to these principles.

Table 11.1 Evolution of sustainable investing

Year	Development
1971	Launch of Pax World Fund, the first socially responsible mutual fund in the United States
1977	Launch of Sullivan Principles
1980s	Widespread disinvestment from South Africa in protest against apartheid
1989	Launch of Valdez Principles (renamed CERES Principles) formed following Exxon Valdez oil spill
1990	Launch of Domini 400 Social Index
1998	Launch of Corporate Governance Code by UK
1999	Launch of Dow Jones Sustainability Indices
2006	Launch of UNPRI
2008	Green bonds issued by World Bank
2009	Launch of sustainable stock exchange (SSE) initiative global dialogues
2015	Launch of UN SDGs
2015	Launch of G20 Energy Efficiency Investor Statement to address the finance for energy efficiency initiatives for G20 policy makers
2015	Launch of Green Infrastructure Investment Coalition to provide a platform of investors, development banks and advisors for countries to be able to tap financing opportunities for green infrastructure
2017	Launch of Climate Action 100+, the largest corporate engagement initiative by investors
2019	Increase in the number of PRI signatories across the globe

Source: UNPRI and other related sources.

Drivers such as client demand for greater transparency, regulatory pressures and materiality aspects of ESG factors play a crucial role in the uptake of responsible investing. However, the degree to which investors sign up for the PRI principles depends upon the power dynamics between the two parties in question. According to Mitchell's et al., 1997 stakeholder salience theory, the different types of power include (1) normative power, in which the stakeholders perceive that integrating ESG issues are associated with enhanced reputation, and (2) utilitarian power, in which the utility aspects of PRI are associated with legitimacy (both legitimacy of the organization and its role in establishing the business case of ESG issues).

Some of the signatories to the PRI include asset managers, investment managers and service providers. The strength of PRI lies in its capability to provide guidance to managers across different asset classes. Some of the ways in which assistance is provided to different investor classes is explicitly mentioned in Table 11.3.

The PRI members engage with companies in accomplishing their Sustainable Development Goal (SDG) commitments and goals. For this, topics of

Table 11.2 Types of sustainable investing

Form of Investing	Specific Characteristics
Impact Investment	Refers to investment that aims to generate social and environmental impact along with a financial return. Typically, these are investments in small companies or projects with clear social goals, providing them with capital they may not otherwise have accessed.
Responsible Investing	Integration of ESG considerations into investment management processes and ownership practices which can impact financial performance in the medium to long term.
Sustainable Investing	Refers to an investment approach which seeks to consider environmental, social and governance issues to generate long-term financial returns. According to DB analysis, 2012, it includes both SRI and ESG forms of investing.
Socially Responsible Investing	Refers to an investment strategy which seeks to consider both financial return and social/environmental good to bring about a positive change utilizing values driven, risks and return screening.
Ethical Investing	Refers to the practice of using one's ethical principles, moral values and religious beliefs as the primary filters for selection of securities.
Green Investing	Refers to an investment strategy which aims to generate environmentally friendly products.
Islamic Finance	Refers to an investment strategy where financial decisions are undertaken based on the provisions made in Islam. It refrains from investment into the sectors/activities which are prohibited in the Islam religion.

Source: Authors' understanding based on UNPRI & Deutsche Bank (2012).

dialogue are chosen based on an internal ESG materiality analysis. The preferred objectives range from encouraging improved ESG disclosure to supporting investment decision choices, strategies and corporate disclosures. Dialogues are initiated between investment managers and the representatives of the companies based on ESG materiality issues. To increase the rigor of discussions, individual dialogues are preferred in which the company is initially contacted via emails. However, apart from individual dialogues, collaborative group discussions are also conducted depending on the situation. The PRI representatives also participate in the annual general meetings (AGMs) through exercising their voting rights. Post-engagement follow-ups are conducted with companies to understand how companies integrate the recommendations provided. One example of this includes the engagement of the European asset management firm Candrium over food safety concerns and public health issues. This engagement has targeted several SDGs.

Table 11.3 Guidance provided by UNPRI

Investor Type	Nature of Guidance Provided
Listed Equity	• Providing guidance on ESG incorporation through active investment strategies such as screening, integration and thematic • Providing guidance on engagement strategies with companies • Providing guidance on ESG integration through passive investing strategies such as fundamental investing, index investing and smart beta
Fixed Income	In addition to the earlier category, providing guidance on thematic investing through bonds in clean energy, infrastructure, energy efficiency and sustainable agriculture
Passive Investing	• Providing strategies on ESG integration in passive investments such as full replication methods and partial replication methods
Private Equity	• Providing strategies on deal sourcing such as identifying material ESG issues in screening and due diligence • Providing guidance in including material ESG issues in investment memorandum and negotiation of investment agreement • Providing guidance on managing ESG issues on onboarding, engagement and monitoring of a portfolio company • Providing strategies on adding value at exit
Hedge Funds	• Providing guidance in conducting due diligence
Infrastructure	• Providing guidance on ESG integration, monitoring and reporting of ESG issues
Real Estate	• Providing guidance in establishing the business case • Help in identifying the material ESG issues

Source: UNPRI, 2016.

Integration techniques

The common techniques used in the modern SRI strategy include ethical negative screening, environmental/social negative screening, positive screening, community and social investing, best-in-class, financially weighted best-in-class, sustainability/climate change themes, constructive engagement, shareholder activism, integrated analysis and norms-based screening.

• **Exclusionary Screening**

Also known as negative screening, certain companies are excluded from a portfolio based on some criteria which may range from traditional moral values and standard norms to international best practices. In value-based exclusions, it may pertain to traditional values such as refraining from tobacco and alcohol companies, and hence the focus is on generating business for the company. In norms-based screening, the focus is on the internationally accepted practices such as human rights and labor standards.

- **Best-in-Class**

This refers to selecting the companies who are best among their peers in terms of better integration of ESG issues.

- **Thematic Investing**

This refers to investment based on themes such as clean tech, green real estate, sustainable forestry, agriculture, education and health. However, this is not restricted to only ESG issues. Examples of this include water- and air-themed funds, alternative energy–themed funds, food and agriculture–themed funds, etc.

- **Impact Investing**

This refers to investing for generating and measuring environmental and social impacts along with a financial return. According to the Global Impact Investing Network (GIIN), the practice of impact investing has four core characteristics: (1) investors intend to have a social and/or an environmental impact; (2) investments are expected to generate a financial return on capital and, at a minimum, a return of capital; (3) investments are to generate returns that range from below-market to risk-adjusted market rate; and (4) investors are committed to measuring and reporting the social and environmental impacts.

- **ESG Integration**

Also known as ESG investing, this refers to systematic integration of ESG issues as factors in investment analysis. However, unlike the best-in-class method, this does not include peer group benchmarking. Simply put, it analyzes the risks and opportunities emanating from ESG factors through inclusion in portfolios.

- **Shareholder Activism**

This refers to entering into dialogue with companies on sustainable issues and exercising ownership rights to influencing change. This implies shareholder activism to impact outcomes and practices on ESG issues. However, this is in sharp contrast to the idea of voting in which investors simply sell off the investments with questionable practices (CFA, 2015). Some of the strongly practiced activism includes voting in AGMs, writing letters to the board of directors of the company, filing shareholder resolutions, meeting with company representatives, filing complaints with the regulator and issuing statements to press and media. In the United States, this is governed by the Dodd-Frank Wall Street Reform and Consumer Protection Act enacted in July 2010. However, the practice of ESG consideration by activist shareholders is not common. There are instances of an increase in ESG shareholder proposals which are usually sponsored by pension funds, ESG-oriented investment managers, religious groups or coalitions of like-minded investors. For equity arbitragers with a myopic vision, ESG integration is still not relevant. Apart from corporate governance, some of the other ESG resolutions filed in the United States in the proxy season include climate change, political activity, human rights and health (Luc Hoffmann Institute, 2018).

USSIF, 2018 has analyzed the growth of sustainable investment strategies over the period 2016–2018. Apart from norms-based screening, other sustainable investment strategies have witnessed a positive rate of increase over the period 2016–2018. There are also regional variations in these strategies across geographies. Although Japan holds 7% of global sustainable investing assets, it holds a major share in corporate engagement and shareholder action strategies. Canada holds a much larger proportion of assets within norms-based screening vis-à-vis other screening strategies. While three-fourths of assets under norms-based screening takes place in Europe, the United States holds the majority of global assets in sustainability investing, impact/community investing, positive/best-in-class investing and ESG integration (USSIF, 2018).

Evaluation techniques

ESG integration involves analysis of all material factors in investment analysis and investment decisions, including ESG factors. ESG evaluation includes analyzing ESG information and its impact on finances, identifying material ESG factors and making investment decisions based on all material factors. The framework for ESG evaluation consists of qualitative analysis, quantitative analysis, investment decision and active ownership assessment. The tools for quantitative assessment vary across buy-side investors and sell-side brokers. According to UNPRI (2016), the following strategies are integrated by buy-side investors.

- **Fundamental Strategies**

  These are used by investors to identify investment opportunities in companies by making assumptions on future performance based on analysis of economic trends, competitive environment, market performance of company products vis-à-vis its peers and quality of the senior management. Some of the fundamental strategies include financial forecasting of financial parameters and firm valuation using valuation models.

o   Financial forecasting – These are conducted through adjustments in the income statement, adjustments in the balance sheet and adjustments in cash flow statements.

o   Company valuation models – These models are used by investment managers to value a firm. These include the dividend discount models, the discounted cash flow model and adjusted present value model. Some of the common measures used to incorporate ESG factors in the company models are calculation of terminal value, adjustment in the beta or discount rates or by conducting a scenario analysis.

- **Quantitative Strategies**

  These strategies for portfolio construction harness data using mathematical models and statistical techniques to outperform their benchmarks. The various stages of this process include analyzing data and statistical testing, building models and back-testing and finally implementing the model.

- **Smart Beta Techniques**

  This is another strategy of portfolio construction in which the constituents of the portfolio are weighed by a factor other than market capitalization such as value, dividend yield, momentum, growth or volatility. This is done to lower the downside risk or increase dividend yield. The smart beta strategies are usually grouped into two categories: heuristic-based weighting methodology in which weights are calculated using simple, heuristic rules which are systematically applied across all constituents and optimization-based weighting methodologies in which complex optimization techniques are used to create portfolios maximizing return or minimizing risk.

- **Passive and Enhanced Passive Strategies**

Passive investment strategies seek to match the performance of a market or a section of a market by closely tracking the return of a capitalization-weighted index. The various ways of undertaking a passive investment strategy include full replication (buying all the constituents of an index), partial replication methodology (investing in a sample set of constituents of an index and adjusting their weights so that the fund matches the index on

Table 11.4 Practical examples of ESG integration

Strategy	Company
Quantitative strategies	Arabesque Asset Management, Auriel Capital
Fundamental strategies	RobecoSAM, RBC Global Asset Management, Allianz Trust Investments
Smart beta strategies	Calvert Investments, AXA Investment Managers
Passive strategies	MSCI, BlackRock, SD-M

Source: KKS Investors.

characteristics such as market capitalization and industry weightings) and using derivatives to track an index.

The purpose of enhanced passive investments is to match the performance of a capitalization-weighted index to either reduce the downside risk relative to a capitalization-weighted index or beat its performance. This is achieved by using the index and its constituent weights as the core of the portfolio and engaging in restricted active strategies, including divesting certain securities, adjusting the weights of constituents and trading derivatives.[2]

Some of the company examples where these strategies have been used by investors are provided in Table 11.4.

Growth of sustainable investment strategies

The investment management industry is in the midst of a fundamental strategy shift due to the increased pertinence of ESG issues and their heightened materiality aspects. It is a known fact today that the risk to return trade-off in investments is affected by ESG issues, and investors are increasingly factoring sustainable factors in their portfolios. ESG considerations are becoming an important driver of long-term investment returns from both an opportunity and a risk mitigation perspective. At the start of 2018, the global sustainable investing market had reached to $30.7 trillion in the five major markets. As evident from Table 11.5, there has been exceptional positive growth of sustainable assets in the countries of Japan and Australia in the period 2014–2016. Japan has continued the trend in the period 2016–2018.

There has been a steady growth of ESG integration in asset classes, especially in Europe, the United States, Japan and Canada. A study by USSIF, 2018 indicates that the highest integration is in public equity followed by fixed income, private equity, venture capital and real estate. Additionally, 7% of integration has also taken place in hedge funds, cash or depository vehicles, commodities and infrastructure, which are categorized as 'other' assets.

Table 11.5 Global growth of sustainable assets in the period 2014–2018

Region	2016	2018	Growth in period 2014–2016	Growth in period 2016–2018
Europe	$12040	$14075	12%	11%
US	$8723	$11995	33%	38%
Japan	$474	$2180	6692%	307%
Canada	$1086	$1699	49%	42%
Australia/New Zealand	$516	$734	248%	46%
Total	$22838	$30683		

Source: Global Sustainable Investment Alliance (2018).

Financing instruments based on sustainable investing

The financial market is steadily witnessing the inclusion of ESG factors in traditional financial instruments.

- **Risk-Sharing Impact Bond**

Also known as pay for success, impact bonds cater to a variety of sectors such as environment and society. These bonds are known as risk sharing because based on demonstrated performance of the project, the investor receives the payment. However, if the desired outcome is not achieved, the investor needs to pay a risk-sharing payment, which means that the investor receives little or no interest.

- **Social Impact Bond**

In the social segment, these bonds are commonly known as social impact bonds (SIBs). It is defined as a contract between the public sector or governing authority and the investor, in which the former pays for better social outcomes in certain areas and passes on part of the savings achieved to investors. The return is contingent on the performance of the service providers in achieving specified social outcomes. However, the riskiness of the bonds is assessed by the fact that if the social outcomes are not achieved, the investor can lose its investment (OECD, 2016).

- **Environmental Impact Bonds**

In the environmental segment, the bonds are known as environmental impact bonds (EIBs). However, the terminology differs across geographies. The functioning is similar to that of an SIB. This is a form of debt financing in which the investors purchase a bond, and the return to investors is conditioned to achieving the environmental outcomes. The major similarity between an EIB and a typical green bond is that the former is used for funding environmentally sustainable projects such as green infrastructure; however, the difference emerges from the fact that the return of an EIB depends on the success of the project undertaken, unlike the green bonds. One example of this is the $25 million bond issued in 2016 by the municipal water board in Washington, DC. The proceeds of the bond were used to fund infrastructure for managing storm water runoff and improving water quality. The return to investors is linked to the performance of green infrastructure. Another example is that of the $13 million infrastructure bond created to fund projects in the flood-prone areas in the city of Atlanta. The bond was structured by the impact advisory firm Quantified Ventures, and the community investment platform Neighborly had partnered to make the offering available to the public.

- **Sustainable Development Bonds**

Also known as the SDG bonds, these are issued by the World Bank to fund social-sector projects in health, education, women empowerment, child development and environmental-sector projects. Recently, the International Bank for Reconstruction and Development (IBRD), one of the member institutions of the World Bank, launched a five-year Global Sustainable Development Bond in support of the SDGs on clean water and sanitation (SDG 6) and life below water (SDG 14). The bond has raised CAD 1.5 billion from institutional investors and has been listed on the Luxembourg Stock Exchange.

- **Green Bonds**

These are bonds issued by various entities to fund environmentally friendly projects. A sub-category of this is the climate bonds which are issued to fund climate change mitigation and adaptation projects. Some of the prominent issuers of green bonds include the European Investment Bank, World Bank and Asian Development Bank (ADB), to name a few. Some of the issuers in India include Yes bank, CLP, Exim Bank and IDBI. The proceeds in India have been mostly used to fund renewable energy projects, energy-efficient projects and low-carbon transport.

- **Water Bonds**

The ADB has issued the water bonds, the proceeds of which will be used to finance water-related projects in the Asia-Pacific region. These will include improving the efficiency of water usage, the management of water resources and the management of effluents from sewage treatment plants. The total worth of bonds issued by ADB since 2010 is $1.5 billion (ADB, 2018). Some successful examples of water projects include the Guizhou Rocky Desertification Area Water Management Project in China and the Integrated Participatory Development and Management of Irrigation Program in Indonesia.

- **Education Support Bonds**

The African Development Bank issued bonds in 2013 to support education-related projects in Africa. Though not much is known about these bonds, it is an effort to support the social sector.

- **JICA Bonds**

Several bonds have been issued by the Japan International Cooperation Agency (JICA) in the period 2015–2016 to support infrastructure development and combat climate change in emerging and developing economies. JICA also issues bonds, the proceeds of which are used to fund the accomplishment of the SDGs.

- **ESG Exchange-Traded Funds**

Exchange-traded funds (ETFs) are primarily index funds which are traded on the stock exchange just like stocks. The first ESG ETF, iShares MSCI USA ESG Select ETF, was launched in 2005 and seeks to track the investment results of an index composed of U.S. companies that have positive ESG characteristics as identified by the index provider. Another example is the NAACP Minority Empowerment ETF, which tracks the Morningstar Minority Empowerment Index designed to provide exposure to U.S. companies with strong racial and ethnic diversity policies in place, empowering employees irrespective of their race or nationality.[3] Morningstar uses a weighting methodology that maximizes exposure to companies with high scores on the NAACP criteria, while maintaining risks and returns similar to the Morningstar U.S. Large-Mid Cap Index.

- **Impact Securitization**

This is a unique model of securitization in which the loans are securitized to finance environmental and social projects, thereby creating an impact in a relatively cheaper manner. One example of this is the loan startup in the United States, Sixup, which provides loans to meritorious students hailing from poor financial backgrounds to attend college and university. Goldman Sachs is the largest lender to the start-up. The idea is once the total amount of lending crosses the $100 million in total assets, the loans will be securitized to broaden the financial base.

Potential challenges to sustainable investing

Despite the ongoing momentum on sustainable investing, significant challenges exist which deter its growth and uptake. The following are the primary challenges associated with sustainable investing.

- Lack of standardized data

One of the most serious shortcomings of sustainable investing is the availability of real-time standardized data. While certain ESG databases such as Bloomberg and Thomson Reuters are available for analysis, standardized and comparable ESG databases are not available. This makes comparison of any form and creation of a standard ESG framework practically impossible.

- Short-termism

Evidence suggests that institutional investors and investment managers, especially in developing countries, suffer from short-sightedness as far as ESG issues are concerned. This is mostly because sustainable investing is a long-term investment strategy, and value addition through ESG factors mostly occurs in the long run. Due to a long payback period and associated uncertainties, many investors are unwilling to invest in sustainable assets.

- Lack of skills and expertise

As sustainable investing is a new investment strategy compared to traditional investing, asset managers in developing countries are not technically equipped to analyze its impact and hence manage it.

- Lack of investment choices

Sustainability products such as bonds, indices, mandates, etc., are limited in developing economies due to underdeveloped capital markets. This leaves investors with fewer choices in terms of investments.

- Lack of awareness and preconceived notions about sustainable investing

Investors are mostly unaware of the value-adding capacity of sustainable investments. More importantly, lack of awareness and collective beliefs about responsible investing are impediments to mainstream integration (Dumas, 2015).

- Lack of proper accounting techniques and valuation

Although some guiding frameworks such as UNPRI are available, investors are mostly clueless on the financial indicators for ESG risk assessments. Financial analysts such as CFA provide some case examples of conducting such analysis (PRI & CFA, 2018).

Recommendations and a way forward

Creation of a viable ESG market, especially in developing countries, is dependent on certain factors.

- Fair perception and awareness about ESG factors by market participants

Financial institutions, stock exchanges, institutional and retail investors, market regulators and companies need to acknowledge and incorporate sector-specific ESG factors into their business portfolios. Awareness and information flow ensure market efficiency of the sustainable financial products.

- Availability of sustainable investment choices in the market

The extent of ESG integration depends on the sustainable investment choices available in the market. These choices may be in the form of sustainable mutual funds, bonds, pension funds, etc. Sinha and Datta (2019) state that the crucial disparity between the sustainable markets of developed and developing economies lies in the availability of investment choices. Developed markets such as Germany, Switzerland and Austria have witnessed a rapid increase in the corporate pension funds, structured products, sustainability indices, bonds and mandates, which provides adequate choices to investors. On the other hand, the investment choices are limited in developing nations due to underdeveloped financial markets.

- Integrating ESG principles in the mission, vision and statement of companies

Market acceptance of ESG investing is a sustained process which will take some time to gain momentum in developing economies. Understanding, recognizing, acknowledging and finally incorporating the ESG factors in business strategies is a complex process. Businesses need to be signatories of the international frameworks such as UNPRI, Carbon Disclosure Project (CDP), Equator Principles, etc., to highlight a few.

- Addressing informational asymmetries between investors and companies

Signaling plays a critical role in the process of information flow between companies and investors. Companies need to be vocal about the ESG efforts undertaken by them. Similarly, investors also need to communicate on the sustainable investment choices available.

- **Creating sustainable investment choices in the debt market**

Several studies have dealt with the impact of ESG factors in stock markets. Generally speaking, the debt market exhibits a considerable weight for sustainable corporate finance, for which creditors should basically play a significant role in the transmission of CSR into valuation of financial instruments (Menz, 2010). However, due to a lack of adequate empirical evidence, inconclusive results relating to the impact of ESG factors on risk premium in the bond market and insufficient data on individual terms of bank loans, etc., this area still remains highly unexplored.

- **Creating sustainable investment choices in the real estate market**

Real estate is one of the underexplored markets in the domain of sustainable corporate finance. In developed economies of Canada and the United States, for companies that invest in and develop commercial real estate assets, the awareness of extra-financial performance of ESG factors have led to significant interest in responsible property investment (Hebb et al., 2010). While some of the institutional investors are integrating ESG standards in their real estate investment portfolios, many are not. Reputation and financial risks are the primary concerns.

- **Role of policy makers and market regulators in facilitating and driving demand for sustainable products**

Policy makers, market regulators and operators play a crucial role in accentuating the ESG market. The framework for environment and corporate governance should be able to incentivize the process rather than creating any impediment to business. Efforts need to be undertaken to fill the existing loopholes in the legal framework. According to GRI, alongside financial data, ESG information is becoming more important in assessing a company's value and provides a more complete picture of its resilience and health. Hence, it is the duty of the market regulators to aid companies in disclosing ESG information.

Notes

1 Accessed at www.dnvgl.com
2 www.unpri.org/listed-equity/esg-integration-in-passive-and-enhanced-passive-strategies/15.article
3 https://hbr.org/2019/01/the-state-of-socially-responsible-investing

References

ADB. (2018). Water Financing Partnership Facility-Annual Work Program. Available at https://www.adb.org/sites/default/files/institutional-document/400796/wfpf-annual-work-program-2018.pdf, last accessed on 09.09.2019.

Benijts, T. (2010). A Framework for Comparing Socially Responsible Investment Markets: An Analysis of the Dutch and Belgian Retail Markets. *Business Ethics: A European Review*, Vol. 19, No. 1, pp. 50–63.

Brundtland, G.H. (1989). Global Change and Our Common Future. *Environment: Science and Policy for Sustainable Development*, Vol. 31, No. 5, pp. 16–43.

Camilleri, M.A. (2015a). Valuing Stakeholder Engagement and Sustainability Reporting. *Corporate Reputation Review*, Vol. 18, No. 3, pp. 210–222.

CFA Institute. (2015). Environmental, Social and Governance Issues in Investing. Available at https://www.cfainstitute.org/-/media/documents/article/position-paper/esg-issues-in-investing-a-guide-for-investment-professionals.ashx, last accessed on 17.02.2020.

Dumas, C. (2015). *The Challenges of Responsible Investment Mainstreaming: Beliefs, Tensions and Paradoxes*. Faculty of Economics and Business Administration, Ghent, Belgium.

Hebb, T., Hamilton, A. & Hachigian, H. (2010). Responsible Property Investing in Canada: Factoring Both Environmental and Social Impacts in the Canadian Real Estate Market. *Journal of Business Ethics*, Vol. 92, pp. 99–115. DOI: 10.1007/s10551-010-0636-5.

Luc Hoffmann Institute. (2018). Shareholder Activism: Standing Up for Sustainability? Available at https://luchoffmanninstitute.org/wp-content/uploads/2018/04/Shareholder-activism-report-.pdf.

Lydenberg, S.D. (2002). Envisioning Socially Responsible Investing. *Journal of Corporate Citizenship*, Vol. 2002, No. 7, pp. 57–77.

Martí-Ballester, C.P. (2015). Investor Reactions to Socially Responsible Investment. *Management Decision*, Vol. 53, No. 3, pp. 571–604.

McCann, L., Solomon, A. & Solomon, J. (2003). Explaining the Growth in UK Socially Responsible Investment. *Journal of General Management*, Vol. 28, No. 4, pp. 15–36.

Menz, K.M. (2010). Corporate Social Responsibility: Is It Rewarded by the Corporate Bond Market? A Critical Note. *Journal of Business Ethics*, Vol. 96, No. 117. https://doi.org/10.1007/s10551-010-0452-y.

Mitchell, R.K., Agle, B.R. & Wood, D.J. (1997). Toward a Theory of Stakeholder Identification and Salience: Defining the Principle of Who and What Really Counts. *Academy of Management Review*, Vol. 22, No. 4, pp. 853–886.

Nilsson, J. (2009). Segmenting Socially Responsible Mutual Fund Investors: The Influence of Financial Return and Social Responsibility. *International Journal of Bank Marketing*, Vol. 27, No. 1, pp. 5–31.

OECD. (2016). Social Impact Bonds: State of Play and Lessons Learnt. Available at https://www.oecd.org/cfe/leed/SIBs-State-Play-Lessons-Final.pdf.

Ogrizek, M. (2002). The Effect of Corporate Social Responsibility on the Branding of Financial Services. *Journal of Financial Services Marketing*, Vol. 6, No. 3, pp. 215–228.

PRI & CFA. (2018). ESG in Equity Analysis and Credit Analysis. Available at https://www.cfainstitute.org/-/media/documents/support/future-finance/esg-integration-overview.ashx?la=en&hash=9D3B70B753958B3B6051B9F4C17A4D927879F71F.

Richardson, B.J. (2008). *Socially Responsible Investment Law: Regulating the Unseen Polluters*. Oxford University Press, Oxford.

Schroders. (2018). Schroders Institutional Investor Study 2018: Institutional Perspectives on Sustainable Investing. Available at https://www.schroders.com/en/sysglobalassets/schroders_ institutional_investor_study_sustainability_report_2018.pdf.

Sievänen, R., Sumelius, J., Islam, Z. & Sell, M. (2012). From Struggle in Responsible Investment to Potential to Improve Global Environmental Governance Through UN PRI. *International Environmental Agreements: Politics, Law & Economics*, Vol. 13, No. 2. DOI: 10.1007/s10784-012-9188-8.

Sinha, R. & Datta, M. (2019). Institutional Investments and Responsible Investing. In: Ziolo, M. & Sergi, B. (eds) *Financing Sustainable Development*. Palgrave Studies in Impact Finance, London.

Sparkes, R. (2003). *Socially Responsible Investment: A Global Revolution*. John Wiley and Sons, Chichester.

UNPRI. (2016). A Practical Guide to ESG Integration for Equity Investing. Available at https://www.unpri.org/download?ac=10, last accessed on 17.02.2020.

UNPRI & Deutsche Bank. (2012). Sustainable Investing: Establishing Long Term Value and Performance. Available at https://www.db.com/cr/de/docs/Sustainable_Investing_2012---Establishing-long-term-value-and-performance.pdf.

USSIF. (2018). Report on US Sustainable, Responsible and Impact Investing Trends. Available at www.ussif.org.

Uzsoki, D. (2020). Sustainable Investing: Shaping the Future and Above. Available at https://www.iisd.org/sites/default/files/publications/sustainable-investing.pdf.

Chapter 12

Sustainable financial systems and a new approach to financial stability

Magdalena Ziolo

Introduction

Financial systems are transforming toward sustainability globally. This is dictated by the experience of the 2008 financial crisis and the impact of non-financial factors on the financial and real spheres. The costs of the impact of nonfinancial activities on the economy and the financial markets, especially those related to the effects of climate change, make it necessary to take urgent measures to transform financial systems in the world so they match sustainable development and achieving goals. Only sustainable financial systems are adequate from the point of view of achieving all sustainable development goals, and, above all, environmental goals, which pose the biggest challenge for both developing and developed countries. The effective functioning of sustainable financial systems requires the provision of a series of activities that condition the fact that the system is sustainable, i.e., activities at the planning, organizing, functioning, and controlling stages of these systems. In particular, the quality of reporting and control plays a key role here, but the decision-making process and environmental, social, and governance (ESG) risk management are not insignificant. Cooperation between the public and private sectors is very important, because without it, integrating the market and public financial system for sustainable development is not possible, and these systems are complementary and interact with each other. It is necessary to change the methodology for assessing financial systems, not only the criteria for evaluation but also the methods of evaluation. Important are the so-called sustainable benchmarks and targets that financial systems strive to achieve. The construction of an integrated measure/index for assessing the sustainable financial system should be considered, which, in addition to economic variables, will consider social and environmental variables, providing a comprehensive view of the financial system as an element conditioning the achievement of economic, social and environmental goals. The rest of the chapter is organized as follows: in the second section theoretical aspects referring to the financial system, financial stability, and sustainability have been presented. The third section presents

ESG factors that impact on a financial system. The fourth section presents challenges and prospects the contemporary financial systems face in sustainability context. The last section includes a conclusion.

The financial system, financial stability, and sustainability

The literature on the subject describes a dual approach to defining the financial system. The first one is the functional approach. The financial system is usually described by pointing to its key activities or functions that it performs in the economy, ignoring institutional and structural differences (Bodie & Merton 1998). The second approach emphasizes the structural and institutional aspects of the financial system (IMF 2006). According to the first approach, a financial system "consists of institutional units and markets that interact, typically in a complex manner, for the purpose of mobilizing funds for investment and providing facilities, including payment systems, for the financing of commercial activity" (IMF 2006). Taking into consideration a functional approach, "the primary function of any financial system is to facilitate the allocation and deployment of economic resources, both across borders and across time, in an uncertain environment" (Bodie & Merton 1998, p. 7). A financial system is commonly defined by its features in the context of stability. In this approach, a financial system is "in a range of stability whenever it is capable of facilitating (rather than impeding) the performance of the economy, and of dissipating financial imbalances that arise endogenously or as a result of significant adverse and anticipated events" (Schinasi 2004). Čihák et al. (2012) proposed analyzing financial systems with regard to the following criteria: depth, access, efficiency, and stability. Verlis (2010) proposed an extended approach based on financial development, financial soundness, or financial vulnerability. A "stable financial system is one that enhances economic performance in many dimensions, whereas an unstable financial system is one that detracts from economic performance" (Schinasi 2004, p. 10). A stable financial system is capable of allocating resources and absorbing shocks efficiently (Schinasi & Schinasi, 2006). A stable financial system guarantees financial stability. Schinasi (2004) discusses five key principles that define financial stability:

- Financial stability takes into consideration the different aspects of finance (and the financial system), among them infrastructure, institutions, and markets.
- Financial stability means that finance adequately fulfills its role in allocating resources and risks; mobilizing savings; and facilitating wealth accumulation, development, and growth; however, it should also assure that the systems of payment throughout the economy function smoothly.

- The concept of financial stability relates not only to the absence of actual financial crises but also to the ability of the financial system to limit, contain, and deal with the emergence of imbalances before they occur.
- Financial stability should be couched in terms of the potential consequences for the real economy.
- Financial stability should be thought of as occurring along a continuum.

The complexity and multidimensionality of financial stability mean that various indicators are used to measure this phenomenon. Table 12.1 presents the measures for the financial sector and financial market commonly used in the literature and financial stability reports.

Financial stability has been included in the Treaty on the Functioning of the European Union (TFEU) as an objective of the European Central Bank (Article 127 V(2)). Activities for the sustainability of financial systems are visible at the level of central banks. It is observed by quantitatively easing and assessing the context of sustainable development on the 2008 financial

Table 12.1 Financial stability indicators based on a literature review and financial stability reports

Scope	Literature Review	Financial Stability Reports
Financial sector	Monetary aggregates	Profitability
	(Real) interest rates	Capital ratios
	Growth in bank credit	Credit (loans)
	Bank leverage ratios,	Liabilities (deposits)
	Nonperforming loans	Liquidity
	Risk premia (CDS):	Credit risk
	Credit risk component of 3-month LIBOR – OIS spreads	Market risk
	Capital adequacy	Interest rate risk
	Liquidity ratio	Asset quality
	Standalone bank credit ratings	Sectoral/regional Systemic focus
	Sectoral/regional concentration, systemic focus	
Financial market	Change in equity indices	Government bonds
	Corporate bond spreads	Corporate bonds
	Market liquidity (government bonds, liquidity risk component of 3m LIBOR – OIS spreads)	Money markets
	Volatility	Equity prices
	House prices	Real estate prices

Source: Author's elaboration based on B. Gadanecz, K. Jayaram: *Measures of financial stability – a review*. A chapter in Proceedings of the IFC Conference on "Measuring financial innovation and its impact", Basel, 26–27 August 2008, 2009, vol. 31, p. 365–380 from Bank for International Settlements.

crisis. On the one hand, this unconventional tool of monetary policy stimulated the economy after the crisis by increasing the liquidity of financial markets. Yet on the other hand, this tool had negative consequences on sustainable development in the form of an inflation impulse and speculative bubbles. Perhaps most importantly, however, the negative impacts on the social pillar of sustainable development led to deepening income inequality (Montecino & Epstein, 2015, p. 1). Ultimately, these negative effects were incomparable to the risk of not responding to the crisis and its potential consequences for the global economy. Financial stability is a widely defined concept. At the macro level it refers to monetary stability and the functioning of the payment system. At the micro level it is linked to market structures (the probability of contagion risk) and financial institutions (Creel et al. 2015). In the micro context, financial stability is interdependent with financial depth, one of the criteria proposed by Čihák et al. (2012). Financial depth "captures the financial sector relative to the economy. It is reflected by the size of banks, other financial institutions, and financial markets in a country, taken together and compared to a measure of economic output" (Čihák et al. 2012; Demirgüç-Kunt et al. 2008, 2011; King & Levine 1993; Levine & Zervos 1988). An extended approach to the assessment of financial systems is based on the financial development sub-index. The index consists of indicators such as the following: market capitalization as a share of gross domestic product (GDP) captures the development of the capital markets, while the ratio of total credit to GDP provides information on the ability of credit institutions to carry out their intermediate functions (Verlis 2010). Financial vulnerability retains the ratio of reserves to deposits and notes and coins to Monetary Aggregate (M2) and acts as an early warning indicator (Verlis 2010). Financial soundness measures the solvency of credit institutions in the financial system (Verlis 2010; Andrés-Alonso et al. 2015). The Principles for Responsible Investment (PRI) point out that the financial system does not operate sustainably and often fails society (PRI 2020). Contemporary discussions about the financial system in this context are about stability and sustainability. The Global Risks Report (2020) indicates that environmental risk is dominant among 10 types of risk in terms of impact. Environmental risk is a key component of the risk of nonfinancial factors known as ESG risk. The term "ESG" refers to nonfinancial factors and has been used since 2004. Back then, under the UN Global Compact's "Who Cares Wins" initiative, attention was paid to the connections and dependencies between ESG factors (Stampe 2014, p. 12). These factors are strongly inscribed and related to sustainable development and, as pointed out by B. Scholten (2006), finance is an important determinant of sustainable development. Therefore, frequent analysis of the relationship between financial and nonfinancial factors should occur (Scholten 2006, pp. 19–33). The impact of nonfinancial factors (ESG) on the financial system is often taken into account by financial institutions (i.e., central banks), as evidenced by the first initiatives concerning the so-called stress climate tests. The Bank of England was the first central bank to conduct comprehensive tests for how well the financial system could absorb risk associated with climate change (Bank of England 2019). The Dutch Central Bank, in turn, analyzed the exposure of the Dutch financial system to carbon-intensive sectors and potential losses related to climate-associated events (Climate risks to European banks 2020). An interesting initiative is the Central Banks and Supervisors Network for Greening the Financial System (NGFS), which is a forum that launched in 2017 to accumulate funding to support the transition toward a sustainable economy and accelerate the greening of the financial system. The European Central Bank monitors the growth dynamics of the sustainable finance market and intends to publish a manual for central banks on ESG best practices in portfolio management. Fatemi and Fooladi (2013) argue that in the very near future, good ESG performance will be a new common standard. In sum, nowadays the financial system must be sustainable in order to be stable. The PRI Initiative defines a "sustainable financial system as a resilient system that contributes to the needs of society by supporting sustainable and equitable economies, while protecting the natural environment". There are different types of sustainable financial systems (Table 12.2).

The financial system may be more environmental, social, or governance oriented. The financial system that is partially ESG oriented is unsustainable. The sustainable financial system is fully ESG oriented. A methodological approach for sustainable financial systems is presented in Table 12.3.

An efficient financial system is a prerequisite for the proper functioning of the economy (real sphere), as it is responsible for both capital allocation and risk management. It is assumed that both of these roles are filled by the financial market in an effective manner. The logic of the financial market has a one-way dimension and assumes the maximization of profit, in particular return on investment (Pisano et al. 2012). In this approach, the concept of sustainable development follows a different approach because maximizing profit is not an overarching goal and an end in itself. Sustainable development is a process that realizes the needs and expectations of the present generation without disturbing or diminishing the ability of future generations to meet their needs based on existing resources, which should be transferred in good condition (WCED 1987). Intergenerational balance in resource management can only be ensured if the management process considers and respects the three pillars (three orders) constituting sustainable development: economic, social, and environmental. Environmental and social aspects are increasingly taken into account as criteria in the process of making financial decisions and conducting risk assessment by financial institutions. Government regulations also play a significant role in the growing role of environmental risk and reputation risk.

Table 12.2 Types of sustainable financial systems

Sustainable financial systems (SFS) typology	The dominant component*	Incorporated ESG** factors	Sustainability pillars***	Risk mitigation	Type of stability	Value****	Perspective
SFS1 unsustainable (soft)	P>M	E	Ec	Financial risk	Financial stability	F, Ec	Short term
SFS2 semi sustainable (mid)	P≥M	E,S	Ec,S	Financial risk	Financial stability	F+Ec+S	Short and mid term
SFS3 sustainable (hard)	P≤M	E,S,G	Ec,E,S	Financial and nonfinancial (ESG) risk	Financial stability and financial sustainability	F+E+Ec+S+G	Long term

* P–public financial system; M–market financial system;
** E–environmental, S–social, G–governance;
*** Ec–Economic, E–Environmental, S–social;
**** F–financial; S–social; E–environmental, Ec–economic; G–governance.

Source: Author's elaboration.

Table 12.3 A methodological approach for sustainable financial systems

Sustainable financial systems (SFS)	Financial stability	Financial depth	Financial development	Financial soundness	Financial vulnerability	Financial fragility	Financial sustainability
Measure	1/ Inflation level; 2/ Ratio of state budget deficit to GDP; 3/ Ratio of current account deficit to GDP; 3/ Real effective exchange rate value increase or deterioration.	1/ Private Credit to GDP; 2/ Total Banking Assets to GDP	1/ Market Capitalization to GDP 2/ Total Credit to GDP 3/ Interest Spread 4/ Herfindahl – Hirschmann Index (HHI)	1/ Nonperforming Loans/Total Loans 2/ Capital/Assets 3/ Z-Score 4/ Liquidity Ratio	1/ Inflation Rate; 2/ General Budget Deficit/Surplus (%GDP) 3/ Current Account Deficit/Surplus (%GDP) 4/ REER (change) 5/ Non-Governmental Credit/Total Credit 6/ Loans (%deposits) 7/ Deposits/M2 ("moving io") (Reserves/Deposits)/(Note & Coins/M2)	1/ Share of credit to households for housing purchases in credit total issued to residents; 2/ Loans issued to non-banks to deposits; 3/ Ratio of total deposits to M2 (broad money);	1/ Sustainable benchmarks; 2/ Climate stress tests; 3/ESG risk-weighted assets; 4/ Green/brown financial flows/assets 5/ Taxonomy for sustainable finance 6/ ESG screening 7/ Capital adequacy (green/brown assets)
SFS1 unsustainable (soft)	*	*	*	*	*	*	*
SFS2 semi sustainable (mid)	**	**	**	**	**	**	**
SFS3 sustainable (hard)	***	***	***	***	***	***	***

Source: Ziolo et al. 2019.

Financial system versus ESG risk

The influence of ESG factors applies in particular to banks that play a key role in ensuring financial stability and security, while at the same time determining the processes of economic development. In the literature it is noted that the development of the concept of corporate social responsibility (CSR) is responsible for the emergence of a new risk category known as ESG and understood as "the potential impact of stakeholders on the company" or, vice versa, "the risk to which the company exposes its stakeholders conducting business activities" (ORSE 2012). According to a different approach, ESG is a new dimension of socially responsible investing (SRI) (Rice et al. 2012). At the same time, it is emphasized that ESG is an external risk that is not determined by the financing structure and yet may affect the entity's financial condition and assets. In addition to financial problems, which ESG risk can determine, it can threaten asset owners and business partners (Inderst 2011). As part of ESG risk, three components are distinguished: environmental, social, and governance risk. ESG risk is inextricably linked to the sustainable finance category and is one of the more significant research currents within this category. On the basis of conducted research, the relationships between ESG risk, its determinants, and the rate of return on the investment portfolio are increasingly emphasized, indicating neutral or positive correlations between variables and the relationship between ESG risk determinants and the cost of capital (e.g., research conducted by Edmans, Li, and Zhang). The ESG risk has an impact on financial risk, in particular through environmental risk on credit risk (Figure 12.1).

Among the environmental risks, climate change is the most influential. ESG issues can have a material impact on a firm's corporate performance and risk profile and on the stability of the financial system (IMF 2019). Climate change requires mitigation measures, on the one hand, and financial markets and systems to adapt to climate change, on the other, and this creates a need for financial capital, instruments, and regulations. Adjustment changes are noticeable on both sides of the financial system, public and market, which offer new mechanisms to counteract the negative effects of change (Krogstrup & Oman 2019). In a 2014 report of the Intergovernmental Panel on Climate Change (IPCC), a chapter dedicated to climate financing was presented for the first time. One of the key postulates of this chapter was that resources for combatting climate change should be significantly increased in subsequent periods in both developed and developing countries (Gupta et al. 2014). Climate risk has a significant impact on financial markets. In particular, global warming affects the valuation of fossil fuels, which may lose value. In turn, more frequent natural disasters determined by climate change can lead to significant losses in asset value and insurance losses (Potential Impact of Climate Change 2019). Therefore, the scale of the impact of climate change on markets and institutions is so large that it can significantly affect the security and stability of financial systems

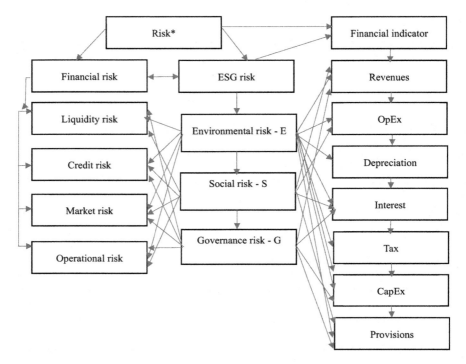

Figure 12.1 The relationship between ESG risk and financial risk

Source: Author's elaboration based on Risk typology based on: Global Risk Report 2020 www.weforum.org/reports/the-global-risks-report-2020 (access: 25.062020) and Risk Management, SMFG 2013, p. 1, www.smfg.co.jp (access: 25.062020); Guidance Note Integrating ESG factors into financial models for infrastructure investments. B Capital Partners, WWF Switzerland 2018, p. 11 http://awsassets.panda.org/downloads/wwf_guid ance_note_infra_.pdf(dostęp: 11.10.2019)

in the long run. A significant risk related to climate change is connected to the fact that in conditions of increasing and more frequent weather threats, the resulting losses will not be insured because the premiums will become too expensive. This increases the indirect risk for the financial market (so-called secondary effects) because uninsured losses can cause a loss of value for companies and a greater risk of loan defaults. In some cases, governments can respond with assistance programs and counteract such losses, which in turn would burden public finances. Indirect risk from uninsured damage, leading to lower creditworthiness by governments, is more likely to exist in poorer, less developed countries (Potential Impact of Climate Change 2019). The consequences of climate change associated with extreme weather events can also cause a direct risk to the financial market. This risk manifests itself in the form of higher and more unstable costs for the insurance sector and has a dimension of potential operational risk related to the closing of bank branches in the event of extreme events (Potential Impact of Climate Change 2019). The consequences of climate change associated with extreme weather events can also cause a direct risk to the financial market. This risk manifests itself in the form of creating higher and unstable costs for the insurance sector and has a dimension of potential operational risk related to the closing of bank branches in the event of extreme events (Potential Impact of Climate Change 2019). In the context of ESG risk, this includes especially negative externalities like climate change and social exclusion. The COVID-19 pandemic confirms arguments about the growing importance of including nonfinancial factors in real and financial spheres. Health risks and more precisely the risk of disease and infection is ranked 10th for what would have the highest impact (Global Risks Report 2019, p. 5). The strength of this type of impact has been confirmed by national and international socioeconomic reports (KPMG 2020; Banking on the Future 2020). The cooperation of banks with entities violating environmental standards creates a risk of losing reputation. The loss of the ability to generate revenues is also determined by the health risk included in the social risk category, which strongly affects the economy and creditworthiness of business entities. This situation is particularly noticed during the pandemic, when actions taken by governments radically influence basic macroeconomic parameters (e.g., GDP, inflation, unemployment, interest rates, deficit and public debt) and the financial situation of monetary and nonmonetary institutions. Recommendations for the financial sector in the form of social and environmental risk management principles in financial institutions, known as the Equator Principles (EP 2013), played a key role in the implementation of sustainable development postulates in the sphere of finance and banking. Environmental and social aspects are increasingly considered as criteria in the process of making financial decisions and conducting risk assessment by financial institutions. Government regulations also play a significant role in the growing role of environmental risk and reputation risk. Governments sanctioned the "polluter pays" principle by imposing financial consequences for those who pollute the environment and do not respect the principles of their protection. The financial dimension of environmental risk and the costs it incurs have contributed to including this category in reporting, among others, quantification through cash flows generated by assets exposed to the risk (it results from the International Accounting Standards or International Financial Reporting Standards) or §510 Basel II and Basel III (Pillar 1), which obliges banks to properly monitor environmental risk in relation to legal repayment collateral. Another good practice is the so-called sustainable stock indices that position companies listed on stock exchanges based on the index, depending on the sum of assessment points of sustainable development implementation, as well as good management practices and business ethics (e.g., Dow Jones Sustainability Index, RESPECT Index, and others). During COVID-19, the activities planned to be implemented as part of sustainable finance have accelerated the field of risk management, reporting, and comprehensive information disclosure, with special attention paid to ESG risk. In the case of the European Union (EU), the search for common solutions to stabilize the finances of EU member states is underway. In view of the independence of the fiscal sphere of EU member states, the potential possibilities of using the European Stability Mechanism (ESM) have been indicated (Europa EU 2020). As a result of COVID-19, it is possible to increase the role of micro- and social finance, instruments that stabilize and limit health risk, as well as fiscal instruments (i.e., public finances in the context of counteracting social exclusion and restoring the budget balance) and public aid instruments. The evolution of sustainable finance will therefore be reflected in business models for financial and nonfinancial institutions. Solutions will be within these systems and, more broadly, financial markets in the field of ESG risk management. Potential sanctions related to noncompliance with the law may affect the financial situation of entrepreneurs as well as the value of assets that constitute the subject of legal repayment collateral. In extreme situations, a decision may be issued to suspend the activity, causing a breach of environmental protection requirements and thus the loss of the ability to generate revenues. Although the banking and financial sector has not been listed among the sectors most affected by the pandemic, COVID-19's impact on financial markets and the wider financial sphere cannot be ignored. One of the sectors suffering from the pandemic is the insurance sector (e.g., payouts, premium refunds, drop in revenues). The sector's ability to react and adapt is more limited than the banking sector, which reacted quickly to the pandemic by offering customers the possibility of suspending loan installments. Meanwhile, as the preliminary calculations showed, the insurance sector could not afford to introduce such solutions (the issue of financial stability of insurers) (Skibińska 2020). Undoubtedly, the impact of the pandemic on the financial sector can be attributed to the acceleration of digitized services and an increase in the number of electronic transactions (e.g., payment cards, mobile devices). Further, one's propensity to save increases during a pandemic. As a result, banks expect an increase in the level of deposits and, at the same time, a decline in lending. The pandemic accelerated the digitization processes for financial products and services, forcing financial institutions to change business models to be more sustainable, both environmentally (digitization reduces the carbon footprint of financial institutions) and socially (the offer of financial products and services better responds to social expectations without violating the interests of financial institutions). The share of noncash payments and electronic transactions (e-commerce) has also increased. At the same time, credit risk has significantly increased by the deteriorating financial situation of enterprises and households at risk of bankruptcy. The increase in indebtedness of public finance-sector entities also affects their creditworthiness, although there is no risk of bankruptcy (legally excluded). The multidirectional impact of COVID-19 and the related health risk on finances has resulted in a number of institutions (public and commercial) taking up solutions for the crisis (the so-called lockout plans). In the context of sustainable finance, these activities mean accelerating the integration of nonfinancial factors into the risk management process much earlier than assumed. Currently, each institution takes this type of risk into account when constructing crisis exit plans. The second effect is the acceleration of disclosing information about transactions affected by ESG risk in market entities in order to comprehensively assess the impact on the real and financial sphere. Recommendations were issued by the European Securities and Markets Authority (ESMA), who referred to the issue of market information disclosure, reporting, and fund management. In particular, issuers should ensure transparency of information about the actual and potential effects of COVID-19. Further, issuers should disclose all relevant information about COVID-19 on their fundamental parameters, forecasts, or financial situation (Regulation No. 596/2014 on market abuse [MAR]).

A sustainable financial system: challenges and prospects

The literature emphasizes that the role of ESG risk and its impact on the financial results of business entities will increase due to the increased importance of factors and aspects such as:

- Intensified interactions and dependencies between climate change and biodiversity and modern technologies that will determine a new approach to investment, risk, and return rates.
- High ESG ratings, which will significantly determine the cost of raising capital, in particular the low cost of obtaining debt financing, which can be a significant competition and threat to financing in the form of municipal bonds, the cost of acquisition of which may be higher and limit access of self-employed units of territorial governments for this source of funding.
- ESG risk-weighted assets that will be reclassified (assets not subject to environmental risk as a result of changes in the environment may increase their exposure to this risk), which will allow shaping new diversification strategies and transaction risk management.

Pisano et al. (2012) draw attention to the following challenges that finances and financial systems face in the context of adapting to sustainable development and, as a result, to ESG risk:

- Adapting the long-term three-dimensional perspective of sustainable development to the short-term perspective of finance focused on profit collection.
- Identifying and including externalities in the process of thinking about finance and financing; finances are to become a mitigating factor for negative externalities.

In the context of these two necessary changes and challenges, there are proposals for solutions (Pisano et al. 2012):

- Internalizing external effects in the process of discounting flows.
- Assigning a long-term horizon to investments in view of the need to maintain a perspective of securing financial capital for the future.
- Gradually replacing financial indicators with sustainability indicators.
- Changing profit management perspectives.

In addition, the Organisation for Economic Co-operation and Development (OECD) in its recommendations indicates system solutions that are necessary to carry out adjustment processes in finance (OECD 2018):

- Improved measurement tools (better indicators and tools to assess sustainability goals).
- Policy reforms to create and provide benefits and incentives for financing and investing in sustainable development.
- Improved communication and information flow between stakeholders to better match supply and demand with sustainable financing of sustainable development goals.

At the same time, the challenge for the financial system, as pointed out by Zorlu (2019), remains to:

- Integrate climate risk into global monitoring of financial stability and stress tests.
- Monitor and valuate financial flows in the field of sustainable financing.
- Develop a taxonomy of noncurrent assets weighted by the impact of ESG factors and action plans for sustainable funding and regional cooperation in this area.

The road map for a sustainable financial system report points to three important initiatives that ensure changes within the financial system toward so-called sustainable financing (WB 2017):

- Market initiatives for the development of joint ventures, such as the Sustainable Banking Network (SBN) and the United Nations Environment Programme Financial Initiative (UNEP FI), thanks to which private and public financial institutions have worked on integrating ESG risk into the business models of financial institutions.
- National initiatives: originally, sustainable development activities resulted from initiatives at the national level, which in many cases were associated with national planning processes to implement climate change policies or other long-term strategic development initiatives.
- International initiatives: activities implemented by the G20, G7, the UN, and the Financial Stability Board (FSB) concerning various aspects of sustainable and green funding while also increasingly involving the private sector. These efforts have been complemented by multilateral development banks (MDBs) and other international financial institutions (IFIs) that promote sustainable financing through, among other methods, adopting sustainable practices in financial activities, up to the introduction of sustainable financial products in the offer of products and services.

At the same time, the report draws attention to the key role of regulations and policies in ensuring sustainable funding, in particular regulated areas, including (WB 2017; WBD2015, 2015):

- Strengthening the role of finance in counteracting negative externalities.
- Supporting the growth of the financial market, including the political framework and standards promoting the issuing of green financial products (i.e., green bonds and securities), the development of new market platforms (i.e., social financing and fintech), and the competitiveness of financial centers.
- Disseminating and promoting market transparency and performance by improving the flow of information on sustainable development through the financial system through guidelines, labeling systems, or mandatory requirements.
- Strengthening risk management by incorporating environmental factors (physical and transformation risk) into prudent supervision of financial institutions, overseeing financial markets, and providing stress tests at the sector and financial system level.
- Facilitating flows and services through investments and loans for priority sectors in terms of their impact on achieving sustainable development

goals, sectoral or funding restrictions, insurance requirements, or the provision of financial services as a means of promoting inclusion and supporting development.

- Clarifying the legal framework, including the responsibilities of fiduciary financial institutions, regarding long-term threats and opportunities (such as climate change).
- Increasing the involvement of financial institutions in ESG activities through codes of conduct and guidelines.

Including regulations and policies for sustainable development is expected to shape a sustainable market financial system, which will ultimately have the following features (WB 2017; WBD2015, 2015):

- It will be an integral part of the process of shaping and implementing policies for sustainable development.
- It will take into account issues related to sustainable development and associated risks.
- It will take into account, measure, estimate, and manage the short- and long-term market associated with sustainable development, considering both transaction and systemic risk.
- It will implement integrated intervention measures to minimize barriers to achieving sustainable development goals.
- It will ensure information transparency, which will become a key feature of the integrated financial reporting system.
- Financial technologies (fintech) and other financial innovation mechanisms will redefine the relationships between financial-sector stakeholders, with a particular focus on sustainable finance.
- Long-term standards for managing nonfinancial factor risk will be adopted, the planning and analysis perspective will be extended, and short-termism will be eliminated.
- A common standard of indicators will be introduced that will be widely used throughout the financial system, and stakeholders will have the expertise to include information on daily operations and formulate long-term strategies for action.

Changes in the market financial system toward sustainability are evolutionary and voluntary, including financial institutions implementing initiatives such as Equator Principles (70% progress), PRI (50% progress), Principles for Sustainable Insurance (20% progress), and the UNEP FI (92% progress) (Connecting Financial 2019). These initiatives have different levels of advancement depending on the country and continent. In connection with initiatives to balance the financial system on the market, there are systematically progressing actions aimed at (a) disclosure of carbon footprint by financial institutions; (b) including ESG factors in decision-making (financial, investment) and risk assessment processes; (c) integrated reporting; (d) stress testing; (e) taking into account so-called fixed assets; and (f) moving from considering threats to sustainable development in favor of seeking business opportunities, which sustainable development provides.

When looking at individual segments of the market financial system, it is worth pointing out trends in banking, insurance, and broadly understood investments that support the shaping of a sustainable system. There is a noticeable trend in the banking sector in which a "sustainable" approach becomes one of the leading strategies for avoiding credit risk and reputation risk. This risk is particularly manifested in defaults, changes in payment schedules, loss of liability insurance, or negative opinions among shareholders or customers. At the same time, the banking sector after the 2008 crisis is one of the sectors subject to the strongest regulatory pressure and also one in which clients place relatively less trust. These factors mean that the liquidity risk in the banking sector is growing due to regulatory requirements regarding capital adequacy, along with the increased competitiveness of para-banking institutions, in particular the so-called fintechs, with which banks have difficulty competing (Deloitte 2020). There are also opportunities for the banking sector, which are associated with the development of initiatives for the inclusive growth and greening of the economy, and these initiatives require not only funding but also specialist services, including advisory services, which banks offer. Banks' involvement in civic initiatives related to, for example, building sustainable agriculture chains, in particular the food economy, as well as social movements opposed to deforestation of forest areas, is noticeable (Banking on climate change 2020).

Climate change is an important and growing factor of instability in the insurance sector, in particular due to potential insurance losses, withdrawal from high-risk markets, and too-high premiums for insurance protection against the risk of nonfinancial factors (e.g., climate change) that customers will not be able to pay. In terms of investing, there is a change in risk profiles, return on equity and credit risk portfolios, and fixed-income portfolios influenced by changes in investor awareness and ESG factors. Nonfinancial issues, such as work, human rights, and community involvement, are increasingly included in the risk profile and return calculation because they have a significant impact on investors' level of risk. The growing awareness of investors who recognize the links between the results of large, diversified investment portfolios and the economy means that more and more investors are making the decision to include ESG in their analysis in all asset classes. Including ESG factors in investment decisions has become one of the fiduciary obligations. Investors feel the impact of regulatory changes and public policies on selected markets, primarily the fuel market, for which it is possible to devalue fossil fuel resources or make them "orphaned" because policies are being introduced to reduce emissions and transform to a low-carbon economy.

Despite numerous initiatives, many areas still require changes and adaptation, and many barriers and limitations are minimized. The key challenges facing the financial market in building a sustainable financial system at a global level include:

- Developing a new approach and methodology for assessing financial systems based on ESG factors.
- Developing ESG risk-weighted assets.
- Implementing integrated reporting, taking into account the impact of ESG factors on financial transactions and operations carried out by financial institutions.
- Undertaking activities in the field of intersectoral and interinstitutional cooperation to ensure sustainable practices in the financial sector.
- Developing a global ESG risk map for the real and financial spheres.
- Developing and monitoring sustainable financial flows.
- Monitoring the processes of greening financial markets.
- Developing a mechanism for as well as analyzing and assessing the effectiveness of sustainable financing.
- Launching a mechanism for reporting by financial institutions on initiatives for sustainable development.
- Launching a reporting mechanism by cooperating entities (clients, suppliers) with institutions operating in the public and market financial system about initiatives for sustainable growth and development.

The development of green bond and palm oil market activities give insight into how voluntary or regulated standardization and certification ensure the transparency and integrity necessary to develop new, sustainable financial assets. The extent to which greening is mainstreamed will largely depend on the extent to which greening initiatives are included in the financial results and the political and regulatory environment. The mechanism for assessing the processes of "balancing" growth and development should be developed by governments to implement a system of incentives based on an analysis of its effects, motivating market participants to transform business models. Financial institutions must begin to fully adapt to sustainable development. This means not only focusing on new, sustainable products, services, and business lines but also taking into account changes in assessing and limiting the impact of nonfinancial factor risk.

There is currently an information gap regarding the stage at which the financial sector is located in various countries in terms of building a sustainable financial system. Innovations at the sectoral and regulatory level mainly concern investment portfolios that generate a carbon footprint and are exposed to the risk of climate change. Document G20 of the Green Finance Study Group calls for systematic measurement of financial "green" flows. This document also indicates the need to track "brown" financial flows to achieve a positive net impact. A voluntary initiative is being promoted to obtain this information directly from financial institutions to close the information gap (Mountford 2020).

The financial sector remains highly diversified in terms of adapting to the needs of sustainable development. Some institutions show a leading position by implementing innovative solutions such as accounting for natural capital, disclosing information related to climate, or issuing green bonds, and at the same time many institutions have not yet implemented the basic principles and rules recommended in the guidelines. Financial institutions face dilemmas related to decisions on management models and goals, while regulators should consider what supportive measures they can take to stimulate collective mainstreaming and the transformation of the financial sphere toward sustainability.

Countries such as France and the Netherlands have made a regulatory effort to accelerate the adjustment of the real and financial spheres to the requirements of sustainable development. France was the first country in Europe obliged (art. 173 French Energy Transition Law) as asset owners and asset managers (including companies, banks, financial intermediaries, institutional investors, asset managers) to report on the impact of physical risk and transient risk determined by climate change on assets that it owns or manages. In the Netherlands, the parliament obligated banks to report on their role and participation in implementing the provisions of the Paris Agreement and to disclose information about the carbon footprint generated by the credit portfolios of Dutch banks. At the same time, based on the example of the Netherlands, financial institutions have launched an accounting partnership on carbon emissions (Challenges in Managing Sustainable Business 2019). The initiative involves 50 financial institutions with a total asset value of $2.9 trillion that have undertaken to account for environmental factors (climate impact) in their investments and loans. Consequently, a common accounting standard for carbon dioxide emissions was adopted. Thanks to the initiative, it will be possible to assess and disclose information on the impact of greenhouse gas emissions created by financial products of influential banks (including the American Amalgamated Bank and the Dutch ASN and Triodos Banks). The standard will allow investors to assess whether their portfolios are compliant with the Paris Agreement and thus whether they contribute to ensuring "financial flows consistent with low greenhouse gas emissions and developing resistance to climate change"; i.e., whether they implement the postulates of the agreement. In the context of monitoring the impact of climate change on the financial sector, the role of rating agencies is strengthening. Rating agencies value the risk of financial institutions, taking into account the effects of this impact, in the case of the S&P methodology in the form of "bad loans", a decrease in asset value, additional "regulatory" costs, and increased reputation risk. Moody's, in turn, pays attention to the impact of technological risk and the environmental factor on the quality of loans granted by banks. The role of technology in the transformation to a green economy is emphasized and the sectors that will gain and lose from this transformation are highlighted, which in turn will affect the risk of banks' loan portfolios (Challenges in Managing Sustainable Business 2019).

Conclusion

Actions taken to build a sustainable financial system are at varying degrees of advancement globally and internationally. It should be emphasized that for many countries, building a sustainable financial system raises difficulties of not only an organizational or financial nature but also political and requires a change in awareness of the perception of the role of the environment and society in the economy. It is particularly difficult to adapt financial systems to the implementation of environmental goals, which means that it is necessary to make decisions about structural changes in the economy, and above all to initiate the decarbonization process. Financial institutions operating in a sustainable financial system differentiate clients and cooperators depending on the type of business and the sector they represent. In extreme situations, entities that do not meet the criteria of sustainable business are excluded from access to financial services until they adapt their activities to the criteria accepted by financial institutions. Such adjustment processes require financial outlays related to the costs of transformation, which for many units (enterprises and households) are too high to be financed from their own funds. Thus, the question arises about the role of the state and the public financial system in the support process for greening the economy and creating sustainable growth and development. Many countries have made the effort to adapt the public financial system in order to increase its role and influence the decisions of market participants, helping to achieve the goals of sustainable development. To this end, a system of incentives and penalties was created for environmentally friendly and environmentally harmful attitudes using the mechanism of environmental taxes as well as penalties and fines. However, this system is highly incomparable and heterogeneous, hence the difficulty in assessing its effectiveness at the global and national level. The public financial system in developing countries plays a key role as a support mechanism for achieving the goals of sustainable development. Financial markets in developing countries are not mature enough and suited to financing sustainable development to support their goals. Full integration of a sustainable market and a public system has been taking place over the years and is mainly characterized by highly developed countries.

Undertaking actions to create "sustainable" financial systems is also visible at the level of central banks. These banks have made an effort to assess the impact of ESG factors (primarily climate change) on financial stability. Some banks have included ESG criteria in their investment decisions. There are also voices of experts in the scope of introducing changes to the perception of capital requirements for greening the economy. Experts are discussing the differentiation of capital requirements at the "green" level, which may mean that banks could, with lower capital requirements, allocate so-called green loans to finance projects that reduce environmental risk in the long run. Among financial institutions and markets there are visible processes of greening and changing business models toward achieving sustainable development goals through sustainable financing. Initiatives for more sustainable financial systems relate to both processes within the organization's so-called green offices and external processes, such as segmentation of customers in terms of fulfilling the criteria of "clean business," introducing ESG risk management mechanisms, basing the decision-making process on ESG criteria, or, broadly understood, building "sustainable value" through business activities at the "core business" level.

References

Arvidsson, Susanne, ed., *Challenges in Managing Sustainable Business: Reporting, Taxation, Ethics and Governance*, Springer, Cham, 2019, pp. 235–237.

Bank of England, *Bank of England to Stress-Test Financial System Against Climate Change Risks*, 2019, www.independent.co.uk/news/business/news/bank-of-eng land-climate-change-stress-tests-financial-system-mark-carney-a9252311.html (access: 5.5.2020).

Banking on Climate Change, Fossil Fuel Finance Report Card, 2019, www.ran.org/ wp-content/uploads/2019/03/Banking_on_Climate_Change_2019_vFINAL1.pdf (access: 20.3.2020).

Banking on the Future: Vision 2020. Deloitte, 2020, https://www2.deloitte.com/ content/dam/Deloitte/in/Documents/financial-services/in-fs-deloitte-banking-col loquium-thoughtpaper-cii.pdf (access: 20.3.2020).

Bodie, Z., R. Merton, *A Conceptual Framework for Analyzing the Financial Environment. The Global Financial System: A Functional Perspective*, 1998, www. researchgate.net/publication/228224831_A_Conceptual_Framework_for_Ana lyzing_the_Financial_Environment.

Čihák, M., A. Demirgüç-Kunt, E. Feyen, R. Levine, *Benchmarking Financial Systems Around the World*, The World Bank, Financial and Private Sector Development Vice Presidency & Development Economics Vice Presidency, Washington, DC, 2012.

Climate Risks to European Banks, www.bruegel.org/2020/02/climate-stress-test/ (access: 20.7.2020).

Connecting Financial System and Sustainable Development: Market Leadership paper, 2016, p. 3 www.unepfi.org/wordpress/wp-content/uploads/2016/11/MKT-LEADERSHIP-REPORT-AW-WEB.pdf (access: 29.12.2019).

Creel, J., P. Hubert, F. Labondance, Financial Stability and Economic Performance, *Economic Modelling* 48(C), 2015, 25–40.

De Andrés-Alonso, P., I. Garcia-Rodriguez, M.E. Romero-Merino, The Dangers of Assessing the Financial Vulnerability of Nonprofits Using Traditional Measures, *Nonprofit Management and Leadership* 25(4), 2015, 371–382.

Deloitte, 2020, https://www2.deloitte.com/cn/en/pages/financial-services/articles/banking-and-capital-markets-impact-covid-19.html (access: 20.7.2020).

Demirgüç-Kunt, A., E. Feyen, R. Levine, *The Evolving Importance of Banks and Securities Markets*, Policy Research Working Paper 5805, World Bank, Washington, DC, 2011.

Demirgüç-Kunt, A., R. Levine, *Finance, Financial Sector Policies, and Long-Run Growth*, M. Spence Growth Commission Background Paper 11, World Bank, Washington, DC, 2008.

EP, 2013, https://equator-principles.com/ (access: 20.7.2020).

Europa EU, 2002, www.esm.europa.eu/content/europe-response-corona-crisis (access: 20.7.2020).

Fatemi, A.M., I.J. Fooladi, Sustainable Finance: A New Paradigm, *Global Finance Journal* 24(2), 2013, 101–113.

Gadanecz, B., K. Jayaram, *Measures of Financial Stability – A Review*. A Chapter in Proceedings of the IFC Conference on "Measuring Financial Innovation and Its Impact", Basel, 26–27 August 2008, 2009, vol. 31, 365–380 from Bank for International Settlements.

The Global Risks Report 2019, https://www.weforum.org/reports/the-global-risks-report-2019 (access: 5.5.2020).

The Global Risks Report 2020, https://www.weforum.org/reports/the-global-risks-report-2020 (access: 5.5.2020).

Guidance Note Integrating ESG Factors into Financial Models for Infrastructure Investments. B Capital Partners, WWF Switzerland, 2018, p. 11, http://awsassets.panda.org/downloads/wwf_guidance_note_infra_.pdf (access: 20.7.2020).

Gupta, S., J. Harnisch, D.C. Barua, et al., Chapter 16-Cross-Cutting Investment and Finance Issues. In: *Climate Change 2014: Mitigation of Climate Change*, Cambridge University Press, New York, 2018.

How to Integrate ESG Risk into Financial Sector's? Operational Risk Management Methods, ORSE, 2012, p. 1, www.orse.org (access: 6.4.2020).

IMF, Overview of the Financial System, IMF, 2006, www.imf.org/external/pubs/ft/fsi/guide/2006/pdf/chp2.pdf (access: 25.5.2020).

IMF 2019, Sustainable Finance, International Monetary Fund. Monetary and Capital Markets Department, www.elibrary.imf.org/view/IMF082/26206-9781498324021/26206-9781498324021/ch06.xml?language=en&redirect=true (access: 20.3.2020).

Inderst, G., Infrastructure as An Asset Class, EIB Papers 2011.

King, R., R. Levine, Finance, Entrepreneurship, and Growth: Theory and Evidence, *Journal of Monetary Economics* 32(3), December 1993, 513–542.

KPMG, 2020, https://home.kpmg/xx/en/home/insights/2020/04/covid-19-impact-and-implications-to-financial-services.html (access: 20.7.2020).

Krogstrup, S., W. Oman, *Macroeconomic and Financial Policies for Climate Change Mitigation: A Review of the Literature*, IMF Working Papers, WP/19/185, 2019,

p. 18, https://www.imf.org/en/Publications/WP/Issues/2019/09/04/Macroeco nomic-and-Financial-Policies-for-Climate-Change-Mitigation-A-Review-of-the-Literature-48612 (access: 28.4.2020).

Levine, R., S. Zervos, Stock Markets, Banks, and Economic Growth, *American Economic Review* 88, 1988, 537–558.

Montecino, J., G. Epstein, Did Quantitative Easing Increase Income Inequality? Institute for New Economic Thinking Working Paper Series No. 28, October 2015, SSRN: http://dx.doi.org/10.2139/ssrn.2692637.

Mountford, H., *Green Vs. Brown: Shifting Thefinance Flows*, http://climatepolicyiniti ative.org/wp-content/uploads/2014/11/Mountford-OECD.pdf/ (access: 05.1.2020).

OECD, Global Outlook on Financing for Sustainable Development 2019, OECD, 2018.

Pisano, U., A. Martinuzzi, B. Bruckner, *The Financial Sector and Sustainable Development: Logics*, Principles and Actors, ESDN Quarterly Report No. 27, December 2012, p. 27.

Potential Impact of Climate Change on Financial Market Stability, https://yoursri. com/news/potential-impact-of-climate-change-on-financial-market-stability (access: 10.12.2019).

PRI, 2020, www.unpri.org/sustainable-markets/sustainable-financial-system (access: 20.3.2020).

Rice, M., R. DiMeo, M. Porter, *Nonprofit Asset Management: Effective Investment Strategies and Oversight*, John Wiley & Sons, New York, 2012.

Risk Management, SMFG 2013, p. 1, www.smfg.co.jp (access: 25.6.2020).

Schinasi, G.J., *Defining Financial Stability*, IMF Working Paper WP/04/187, October 2004.

Schinasi, G.J., D. Schinasi, *Safeguarding Financial Stability: Theory and Practice*, IMF, 2006, www.imf.org (access: 28.4.2020).

Scholten, B., Finance as a Driver of Corporate Social Responsibility, *Journal of Business Ethics* 68(1), September 2006.

Skibińska, R., *Koronawirus zaszkodzi ubezpieczeniom*, www.obserwatorfinansowy. pl/tematyka/rynki-finansowe/bankowosc/koronawirus-zaszkodzi-ubezpieczen iom/ (access: 28.4.2020).

Verlis, C., *Measuring and Forecasting Financial Stability: The Composition of an Aggregate Financial Stability Index for Jamaica*, Morris Financial Stability Department Bank of Jamaica, 2010, https://scholar.google.com/scholar_loo kup?title=Measuring+and+Forecasting+Financial+Stability:+The+Composition +of+an+Aggregate+Financial+Stability+Index+for+Jamaica&author=Morris,+ Verlis&publication_year=2010 (access: 28.4.2020).

WB, Roadmap for a Sustainable Financial System. A UN Environment – World Bank Group Initiative, 2017, http://documents.worldbank.org/curated/en/90360 1510548466486/pdf/121283-12-11-2017-15-33-33-RoadmapforaSustainableFi nancialSystem.pdf (access: 28.12.2019).

WBD2015, World Bank Document, *Towards a Sustainable Financial System in Indonesia*, The Association for Sustainable and Responsible Investment in Asia (ASrIA), 2015 https://www.ifc.org/wps/wcm/connect/topics_ext_content/ifc_ external_corporate_site/climate+business/resources/towards+a+sustainable+finan cial+system+in+indonesia (access: 5.5.2020).

WCED 1987, The World Commission on Environment and Development, Report of the World Commission on Environment and Development Our Common Future United Nations 1987, https://sustainabledevelopment.un.org/milestones/wced (access: 5.5.2020).

Wildlife Fund, *Environmental, Social and Governance Integration for Banks: A Guide to Starting Implementation*, ed. J. Stampe, Gland, Switzerland, 2014.

Ziolo, M., B. Filipiak, I. Bąk, K. Cheba, How to Design More Sustainable Financial Systems: The Roles of Environmental, Social, and Governance Factors in the Decision-Making Process, *Sustainability* 11, 2019, 5604.

Zorlu, P., *Transforming the Financial System for Delivering Sustainable Development – A High-Level Overview*, Finance Taskforce, IGES, IGES Discussion Paper, 2018, https://pdfs.semanticscholar.org/ea29/f6b6f64b7b2f05df62ea13805a4e0 eec3f75.pdf (access: 3.6.2019).

Chapter 13

Sustainable audit, control and monitoring of Indian companies

Ria Sinha and Manipadma Datta

Introduction

Since the inception of the Sustainable Development Goals (SDGs) in 2015, the private- and public-sector companies in India are finding out ways to incorporate sustainability issues into business operations and strategies. While the underlying drivers may pertain to regulatory pressure, reputational concerns, willingness to improve relations with national governments and litigation risks, businesses are gradually realising the opportunities that might arise from competitive positioning (Pillai et al., 2017), but the adoption of sustainable business practices by private companies is quite a challenging process (Scheyvens et al., 2016). Hence, leveraging the existing risks by using SDG as a tool is a well-acknowledged notion to proliferate business activities and innovation. While investors have graduated from developing environmental, social and governance (ESG) integration approaches across different levels of management, research and applications, the extent of integration varies across economies, and the need for an effective ESG management system has been felt across all investment classes (Sinha and Datta, 2019). According to SDG Fund, 2016, SDG engagement by private businesses may require access to innovation capital, complementary expertise and public finance to leverage upon. Pattanaro and Donato (2018) have analysed the business opportunities arising out of SDGs to the tourism sector. According to them, SDGs 12 and 17 are extremely relevant for several sectors and the tourism sector in particular. Pedersen (2018) states that with the SDGs in place, businesses now have a much clearer set of long-term priorities and a wider scope of engagement with the policy makers and civil society. He is of the opinion that 17 SDGs along with 169 specific targets represent a long-term political framework for businesses to contribute to sustainable development. It is also a framework which helps in synchronising the market expectations with the expectations from society in terms of resource use. Hence, the SDGs empower the business sector to have a long-term vision vis-a-vis seeking newer investment opportunities. Similarly, for those companies that do not possess fitting SDG solutions, the

SDGs provide a guiding framework to reduce risks and costs and assist in business model transformations. There are several estimates on the potential value addition and economic benefits by SDGs. According to a report by the Business & Sustainable Development Commission, the potential economic reward from delivering solutions to the SDGs could be worth at least $12 trillion each year in market opportunities and generate up to 380 million new jobs by 2030.

However, the extent of SDG implementation will depend on the robustness of policy and regulatory frameworks, sustainability accounting tools, compliance mechanisms and credibility of sustainability disclosure reports. The present chapter outlines the various SDG control and monitoring mechanisms in place and emphasises the need for external auditing of sustainability reports which are at present conducted in an arbitrarily manner.

The rest of the chapter is organized as follows. The next section states the importance of sustainability assessments, audits, control and monitoring in general. The tools for these mechanisms are elaborated in the third section. The fourth and fifth sections emphasise the institutional framework of sustainability control, monitoring mechanisms and tools in India. The challenges are highlighted in the sixth section, and the last section provides the structural recommendations.

Importance of sustainability assessments, audits, control and monitoring

Sustainability assessments and monitoring

While this is an emerging field, sustainability assessment (SA) according to Bond et al. (2012) is an early practice to accommodate new situations and contexts. Kates et al. (2001) mentions that SA enables decision makers to evaluate a global-to-local integrated nature and society systems both for the short and long term to determine the actions to be undertaken in order to make a society sustainable. However, the use of this concept has been expanded to include firms as well.

Impact assessments based on sustainability are gradually being used by companies to assess potential risks emanating from sustainability issues and SDGs. The International Association for Impact Assessment defines impact assessment as 'the process of identifying future consequences of a current or proposed action'. According to Waas et al. (2014), there can be two approaches to SA. One is a generic approach which considers an environmental impact assessment (EIA), strategic environmental assessment, health impact assessment and risk assessment, and the other involves more advanced forms of SA which are less clearly defined. Bond et al. (2012) considers SA a 'recent framing of impact assessment' which can also be considered a 'third generation' impact assessment following the EIA and strategic environmental assessment. According to other academic literature, SA is the umbrella term which includes indicator development, product-related assessment and integrated assessments such as EIA (Hacking and Guthrie, 2008; Ness et al., 2007).

Sustainability monitoring forms an integral part of the broad umbrella of SA. Monitoring tools help in identifying the deviations from the proposed plan and implementation gaps. Hence, it helps in spelling out the additional steps required for unanticipated impacts or when those larger than projections occur (Joseph et al., 2019).

With the recent formulation of the SDGs, there is an increasing trend towards the use of the global goals as a guideline for indicator selection in life cycle assessments. In their innovative approach, Wulf et al. (2018) conducts a Life Cycle Sustainability Assessment (LCSA) of electrolytic hydrogen production through the application of preselected LCSA indicators mapped against SDG goals and an indicator framework. The results of the study indicate that 14 out of the 17 goals can be mapped against the preselected LCSA indicators. The analysis shows meaningful differences between the goal-based and the indicator-based assessment. Only the goal-based indicator set comprises all dimensions of sustainability.

Sustainability accounting, reporting and audits

The terms sustainability accounting, reporting and auditing are closely intertwined in business literature, the roots of which emanate from traditional accounting. According to Wells, 1978; Fleischman and Tyson, 1998, financial and cost accounting were designed to meet the needs of external reporting. On the other hand, management accounting techniques were developed to meet the needs of internal reporting, i.e. meeting the needs of managers for relevant data to facilitate decision-making, planning, control and monitoring (Burritt, 2002). The term sustainability reporting which emerged later emphasised the needs of linking traditional accounting with sustainability issues (Gray, 1992; Schaltegger and Sturm, 1992; Mathews, 1997) as well as financial issues. According to Burritt and Schaltegger (2010), there are primarily three approaches followed in sustainability accounting: inside-out, outside-in and a combination of both. In the inside-out approach managers require relevant and reliable information to undertake decisions and hence require a set of pragmatic tools which contribute to the business solutions of environmental and social problems. These tools constitute a pertinent part of systematic, effective and efficient problem solving. Hence, this approach requires development of the existing skill set of accountants to equip them to account for sustainability by building on the traditional accounting approaches (Schaltegger and Burritt, 2000). Some of the tools used in this approach are the sustainability balanced scorecard (Kaplan and Norton, 1992; 2004a, 2004b), eco-control or sustainability management control. These tools help in condensing relevant sustainability issues into key performance indicators and information requirements.

The outside-in approach is another way of considering how management can contribute to sustainable development through reporting for external stakeholders. This can include issues pertaining to environmental, social and corporate governance. While the gamut of these issues is vast, there are no universally accepted indicators. Although certain international frameworks might exist, these issues are usually the ones which have an impact on the overall financial system. According to Schaltegger and Wagner (2006) the outside-in approach consists of stakeholder dialogues and discussions, reports and communicates the corporate contribution of these issues and thereby defines measurement and management activities based on these issues for communicating to various stakeholders such as rating agencies, media, investors and other relevant stakeholders. It is anticipated that external stakeholders have a greater impact on corporate behaviour and performance, compared to internal goals and managerial commitment Vitale et al. (2019).

The fundamental difference between the two approaches lies in the type of stakeholders involved in the process and the extent to which they influence decision-making. One of the internationally recognised reporting guidelines used by organisations include the Global Reporting Initiative (GRI) guidelines, which are based on a triple bottom-line (TBL) approach. Others include environmental and social reports (Bennett and James, 1999). The third approach is a twin-track approach which considers both outside-in and inside-out together (Schaltegger and Wagner, 2006). Tools such as eco-control are based on this approach (Henri and Journeault, 2010).

There are also alternative opinions on the development of sustainable accounting. Fagerström and Cunningham (2016) mentions sustainability audits and accounting are a gradual evolution from sustainability reporting in which an accounting system accumulates information in a systematic manner. These are now used as a risk assessment tool based on the perspective of all stakeholders: financial, social, environmental and technological (Fagerström and Hartwig, 2016). The importance of sustainability accounting emanates from the fact that traditional accounting methods fail to address sustainability issues which can lead to financial distress. However, according to these authors, the postulates of sustainability accounting are developed from the postulates of traditional financial accounting and should possess the elements of business continuity to meet the sustainability objectives, including all activities which impact sustainability and including the reporting period covering the product life cycle starting from extraction of raw materials to waste disposal and possess defined sustainability indicators. Some of the well-known sustainability accounting standards are provided by the Sustainability Accounting Standards Board (SASB) and German Environmental Economic Accounting (GEEA).

Companies usually disclose sustainability information in the form of sustainability reports. According to Kolk (2004), a sustainability report is an entity report that gives information about its economic, social, environmental and governance performance. GRI defines sustainability reporting as 'the practice of measuring, disclosing and being accountable to internal and external stakeholders for organizational performance towards the goal of sustainable development'. According to Godha and Jain (2015), sustainability reporting is a process that facilitates in goal setting and measuring performance in an organisation. It provides a key platform for communicating an organisation's sustainability performance, reflecting positive and negative impacts. The benefits of sustainability reporting include enhanced business performance through reduction of operating costs and improving efficiency, innovating new products and services and improving brand value through product differentiation and integrity management (Venning and Higgins, 2001). According to a report by KPMG (2015), more than 95% of the 250 world's largest corporations publish sustainability reports. To establish credibility of these reports, third-party assurance is conducted by companies worldwide using two recognized professional standards: (i) the AA 1000 Assurance Standard – developed by the Institute for Social and Ethical Accountability – and (ii) the International Standard on Assurance Engagements (ISAE) 3000 Assurance Standard – provided by the International Audit and Assurance Standard Board. Similarly, for social reporting, ISO26000 is a recognised framework. Apart from the GRI and ISO26000, the other guidance providers of sustainability reporting include the Organisation for Economic Co-operation and Development (OECD) and United Nations Global Compact (UNGC).

Tools for sustainability accounting, audits and monitoring

- **Environmental accounting**

This is considered be a sub-category of sustainability accounting. Emanating from the concept of traditional accounting, environmental accounting information can consist of monetary and non-monetary aspects and is relevant for both internal and external stakeholders of the organisation. While the monetary aspects are included in the financial aspects of environmental accounting such as integration of environment-related information into financial data, like earnings and expenses for environment-related investments or environmental liability, the non-monetary aspects comprise the management-related aspects of the environment involving life cycle costing, full cost accounting, benefits assessment and strategic planning.

While this is gradually gaining recognition, the international casual footwear German brand PUMA has been conducting environmental accounting since 2010. The company first published its total economic valuation of $145 million, out of which $51 million is caused by land use change for the production of raw materials, air pollution and waste along its value chain. This $51 million derived from PUMA's environmental profit and loss accounts (E P&L) indicates the immense value of nature's resources and services that are currently taken for granted; however, without these services companies could not sustain themselves.[1] E P&L is a very powerful tool to gauge the environmental footprint of any company and thereby measures which can be taken to reduce the footprint.

- Carbon accounting

Considered a subset of sustainability and environmental accounting, this particular concept relates to accounting for climate change. However, there is no consensual definition of carbon accounting. The other variants of this term are carbon emissions accounting and carbon offset accounting. According to Stechemesser (2012), one of the advantages of carbon accounting is that it can be used to measure (i) carbon emissions, and removal and retain an ongoing inventory of operations-based emissions and (ii) financial statements of impacts resulting from an entity's carbon regulatory environment and transacting strategies. There have been specific applications of carbon accounting in India; however, the scope remains limited.

- SDG Compass

Launched by the UN at an event in 2015, the SDG Compass was jointly developed by the GRI, UNGC and World Business Council for Sustainable Development and is used for the purpose of SDG monitoring. The compass provides guidance to companies on how they can align their strategies and measure and manage their contribution to the realisation of SDGs.[2] It consists of a list of business tools applied by investors and companies worldwide and filters the tools based on the goals.

Institutional framework for aligning towards SDG governance for Indian companies

With the adoption of the 2030 Agenda for Sustainable Development in 2015, the world has agreed to act upon the challenge for accelerating action and achieve the objectives of sustainable development for all. Since then the

Indian government has aligned its efforts with the Indian policy think tank NITI Aayog (National Institution for Transforming India) spearheading the efforts to implement SDGs through cooperative federalism by fostering the involvement of Indian state governments in the economic policy-making process using a bottom-up approach. NITI Aayog released the SDG India index in December 2019 which comprehensively documents the progress made by India's states and union territories (UTs) towards achieving the 2030 SDG targets. The index has been developed in collaboration with the Ministry of Statistics and Programme Implementation (MoSPI), United Nations in India and Global Green Growth Institute (GGGI). The index tracks progress of all states and UTs on 100 indicators drawn from the MoSPI's National Indicator Framework (NIF). The process selecting these indicators had included multiple consultations with union ministries/departments and states/UTs. The SDG India Index report has a new section on profiles of all 37 states and UTs, which will be very useful to analyse their performance on all goals in a lucid manner. The progress of the states on goal implementation is tracked through the SDG dashboard monitored by NITI Aayog. For calculation of the index scores, a composite score has been computed in the range of 0 to 100 for each state/UT based on its aggregate performance across 16 SDGs, indicating the average performance of every state/UT towards achieving 16 SDGs and their respective targets. Accordingly, if a state/UT achieves a score of 100, it signifies it has achieved the 2030 national targets. The higher the score of a state/UT, the closer it is towards achieving the targets. Additionally, the scores have been categorised in the following manner: Aspirant: 0–49; Performer: 50–64; Front Runner: 65–99; Achiever: 100.[3]

Apart from NITI Aayog, the regulatory framework of SDG implementation in India comprises the Ministry of Corporate Affairs (MCA) and line ministries. On the corporate side, the Securities and Exchange Board of India (SEBI) and the National and Bombay Stock Exchanges (BSE, NSE) play a cogent role. Figure 13.1 illustrates the various developments which have taken place so far to rigorously implement SDGs in India.

Sustainability actions of Indian companies

Indian privately listed and public-sector companies have aligned their business strategies to leverage the business opportunities posed by SDGs. While this is imperative, the theoretical foundation lies in the fact that SDGs create significant value, a finding which is corroborated by several social scientists and researchers. Sustainable value is created through systems thinking, systems innovations and reorientation of business models. According to Bocken et al. (2014), business model innovations for sustainability create significant positive impacts and reduce negative impacts

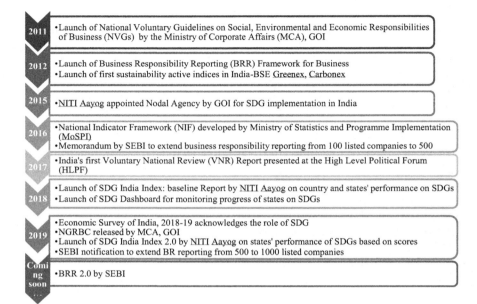

Figure 13.1 Chronology of events for SDG implementation in India
Source: Author.

for the environment through changes in the way the organisation and its value network create, deliver value and capture value or change their value propositions. Sustainability value is created by focusing on the TBL and accounting for improvement in the life cycle of a product through innovative strategies (Fearne et al., 2012).

However, the extent of materiality of sustainability issues varies significantly across industry type, the business model of companies and local conditions (Schoenmaker and Schramade, 2019). Indian companies have resorted to several ways of internalizing SDGs into their business strategies and actions. The following sub-sections elaborate on this.

Aligning sustainability actions with SDGs

Several Indian companies have adopted an inside-out strategy for mapping SDGs in which the company actions are mapped against the SDGs being addressed by them. One of the guiding frameworks adhered by companies in the mining sector is Mapping Mining to the SDGs: An Atlas[4] developed by United Nations Development Programme (UNDP). The Indian company National Mineral Development Corporation (NMDC) has mapped its actions on SDGs and have addressed all 17. Similarly, the TATA group of companies – TATA Chemicals, TATA Steel, TATA Power – have also undertaken SDG mapping of companies. TATA Chemicals has gone a step further to map its corporate social responsibility (CSR) actions against the SDGs.

Business responsibility and integrated reporting

Business responsibility reporting (BRR) is mandatory for Indian listed companies as mandated by the SEBI in 2016 for non-financial disclosures. The coverage has increased from 500 companies in 2016 to 1000 companies in 2019. The template for these reports has a suggestive framework comprising 44 questions and is divided into five sections, which includes data on aspects such as ethics, accountability and transparency; product life cycle sustainability; employees' wellbeing; stakeholder engagement; human rights; environment; public advocacy; inclusive growth; and customer value. More importantly, the guidelines of the BRR framework have been prepared by consulting existing international frameworks and national consultations with business, civil society and academia. The importance of this framework lies in the fact that it adheres to the National Voluntary Guidelines (NVGs) on Social, Environmental and Economic Responsibilities of Business launched in 2011 comprising nine principles and clearly establishes the business case for sustainability. Apart from the BRR, several Indian companies are voluntarily disclosing sustainability information using GRI guidelines. In addition, some multinational Indian conglomerates like Larsen & Toubro (L&T) are conducting integrated reporting. The importance of integrated reporting is ascertained by the fact that it helps to communicate how value-relevant information fits into operations and business strategies, thereby facilitating capital allocation decisions. This is extremely beneficial to a wide range of stakeholders who take informed decisions on their portfolios.

Materiality assessments

A large number of Indian listed companies are undertaking materiality assessments in which a primary evaluation of company actions is conducted through measuring the impact of the actions on long-term financial performance vis-à-vis the pertinence of those actions to stakeholders. A materiality matrix is thereby developed which indicates the actions that are not only relevant for the company but also relevant to the stakeholders. Some of the companies which conduct such assessments are the TATA Group of companies, ASEA Brown Boveri and Mahindra & Mahindra, to name a few.

Sustainability control, auditing and monitoring of Indian companies

Environmental accounting

The practice of environmental accounting is not new in India. However, the effectiveness and transparency vary across sectors and companies. A study by Verma (2002) on the environmental accounting practices of six Indian companies, namely Dr Reddy's Lab, Ranbaxy Laboratories, Balsmapur Chini Mills, Gujrat Ambuja Cement, Hindustan Lever Ltd. and Slaw Wallace Group for the year 2001–02 reveals that the companies made policy statements in a director's report but did not disclose any quantitative figures on expenditures incurred on targets set and achieved with respect to natural resources. Similarly, Bhatt (2005) studied the environmental accounting practices of three companies: Steel Authority of India Limited, TATA Steel and Grasim Industries in the period 2003–03 and concluded that although these companies have taken steps in the right direction, they did not make any effort to calculate the environmental footprint of their activities across operations and the supply chain. Although practices have evolved over the years, environmental accounting is still not conducted in a rigorous and systematic manner in India. Commenting upon sustainability accounting, it's still not fully comprehended by companies and hence is limited to environmental accounting practices only.

Sustainability reports and audits

Mandated by the Companies Act of 2013, CSR reporting is mandatory in India. Social sectors such as health and education are some of the areas emphasised. However, a verification framework for these reports is not yet developed. As observed by Sharma (n.d.), third-party assurance of corporate responsibility reports is accepted by a large number of sustainability reporters in India. According to Kumar and Prakash (2019), the quality of disclosure in CSR reporting is still questionable and needs improvement. In fact, a majority of these reports are assured by using the ISAE3000 standard.

Similarly, several Indian listed companies are voluntarily producing sustainability reports using the GRI G.4 reporting framework. Several companies in India have begun to adopt the AA 1000 Assurance Standard for sustainability reporting. Goel and Misra (2017) state that although CSR reporting by Indian companies has witnessed an upsurge in the recent years, sustainability reports are still in the nascent stage. As far as auditing of these reports is concerned, the reports are either self-assessed, GRI-checked or checked by a third-party independent agency. Consulting firms such as KPMG and Ernst & Young are engaged as external auditors in independently assessing and auditing sustainability reports of listed companies based on GRI reporting. As observed by Sharma (n.d.), companies disclose their corporate governance information in the annual reports; however, few companies have also attempted to link a sustainability framework to corporate governance, and fewer companies have integrated sustainability elements in their risk management framework. The standards adhered include ISO 9001, ISO 14001 and OHSAS 18001 management standards. As rightly examined by the author, although financial reports are subject to compulsory statutory audits, internal audits for non-financial information are not conducted by many companies.

On the regulatory side, the primary public institution which is responsible for conducting audits for central and state governments and government-affiliated entities on SDG matters is the Comptroller and Auditor General of India (CAG). The regulatory framework for such audits comprises policies such as the National Environment Policy, 2006; National Conservation Strategy and Policy Statement on Environment and Development, 1992; Policy Statement for Abatement of Pollution; and the National Forest Policy, 2006. The institutional authority for conducting the audits encompasses the Ministry of Environments, Forests and Climate Change (MOEFCC), Central Pollution Control Board (CPCB), NITI Aayog and the State Pollution Control Boards (SPCBs).

Sustainable monitoring

- NIF Framework[5]

Developed by the Ministry of Statistics and Programme Implementation (MoSPI), this framework provides a list of 306 statistical indicators based on the 17 SDGs and 169 targets. The framework is considered the backbone of monitoring SDGs at the national level which is expected to provide appropriate direction to policy makers and implementors of various schemes and programmes. However, one of the primary gaps of the NIF is that it still requires validation from the private business sector.

- SDG India Index[6]

As elucidated in Section 3, the SDG India index is constructed using 100 indicators and covers 54 targets across 16 goals, barring Goal 17, which primarily focuses on partnerships. While 68 out of 100 indicators are directly taken from the NIF, 20 NIF indicators have been modified or refined for the sake of data availability across all states/ UTs. Based on a robust methodology, some of the frontrunner states as per index scores include Kerala, Himachal Pradesh, Andhra Pradesh, Tamil Nadu, Telangana, Karnataka, Goa and Sikkim. However, the index suffers from certain limitation such as non-representation of Goal 17 and non-inclusion of all the NIF indicators.

Challenges

Several challenges are faced by Indian companies in terms of sustainability reporting. Being voluntary in nature, there is no uniformly accepted standards in publishing the reports and verifying them. This aside, data insufficiency and lack of a centralised sustainability database prevent managers from self-assessing non-financial information. Inadequate sustainability risk assessments by companies also lead to discrepancies. Moreover, the established frameworks cater to listed companies, and small and medium-scale industries (SMESs) are out of the purview and pose a burdensome affair for them. According to Jones et al. (2017), several companies in the tourism sector fail to prioritise the SDGs to be addressed.

Recommendations for increased compliance

The recommendations provided in this section cater to both companies and the regulator.

Recommendations for companies

- Companies need to frame sustainability assessment and monitoring tools for addressing the right SDGs. As already established in this chapter, several companies are conducting sustainability assessments, materiality matrices and auditing their sustainability reports, but the approach isn't uniform. Hence, first of all, there has to be a well-defined established sustainability framework within companies. For this, the international guideline provided by the UNGC can be adopted. One of the approaches is illustrated in Figure 13.2.

Policy recommendations

- Regulators have undertaken several piece-meal efforts to improve compliance. However, national frameworks such as the NIF need to be integrated with the business and private sector to make them user friendly.
- There is a strong need for conducting a compulsory audit of the sustainability reports using internationally recognised agencies.
- There is also a need to define the extent of sustainability issues which can be integrated for accounting purposes. A materiality assessment of key sustainability issues can be conducted for this purpose.
- Reports such as CSR, BRR and corporate governance internal and external audits need to be mandatory.
- For disclosure of sustainability information, uniformity in assurance standards needs to be established.

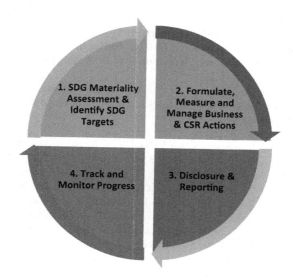

Figure 13.2 Sustainability assessment framework for Indian companies
Source: Author.

Notes

1 https://about.puma.com/en/newsroom/corporate-news/2011/11-16-11-first-envi
   ronmental-profit-and-loss
2 https://sdgcompass.org/
3 https://pib.gov.in/PressReleaseIframePage.aspx?PRID=1597981
4 Available at www.undp.org/content/undp/en/home/librarypage/poverty-reduc
   tion/mapping-mining-to-the-sdgs--an-atlas.html
5 www.mospi.gov.in/national-indicator-framework
6 https://niti.gov.in/sdg-india-index-dashboard-2019-20

References

Bennett, M. and James, P. (1999). *Sustainable Measures: Evaluating and Reporting on Environmental and Social Performance*. Sheffield. Greenleaf Publishing.

Bocken, N.M.P., Short, S., Rana, P. and Evans, S. (2014). A Literature and Practice Review to Develop Sustainable Business Models Archetypes. *Journal of Cleaner Production*, 65, pp 42–56.

Bond, A., Morrison-Saunders, A. and Pope, J. (2012). Sustainability Assessment: The State of the Art. *Impact Assessment and Project Appraisal*, 30, pp 53–62.

Burritt, R.L. (2002). Environmental Reporting in Australia: Current Practices and Issues for the Future. *Business Strategy and the Environment*, 11(6), pp 391–406.

Burritt, R.L. and Schaltegger, S. (2010). Sustainability Accounting and Reporting: Fad or Trend? *Accounting, Auditing & Accountability Journal*, 23(7), pp 829–846. https://doi.org/10.1108/09513571011080144.

Fagerström, A. and Cunningham, G. (2016). Accounting and Auditing of Sustainability: Sustainable Indicator Accounting, SIA. *Sustainability: The Journal of Record*, 12(3), pp 158–162.

Fagerström, A. and Hartwig, G. (2016). Accounting and Auditing of Sustainability: A Modelnter Title. https://en.wikipedia.org/w/index.php?title=Sustainability_accounting&oldid=954011728.

Fearne, A., Garcia, M., Marian, D. and Benjamin. (2012). Dimensions of Sustainable Value Chains: Implications for Value Chain Analysis. *Supply Chain Management*, 17(6), pp 575–581. ISSN 1359-8546. doi:10.1108/13598541211269193.

Fleischman, R.K. and Tyson, T.N. (1998). The Evolution of Standard Costing in the UK and US: from Decision Making to Control. *Abacus*, 34(1), pp. 92–119.

Godha, A. and Jain, P. (2015). Sustainability Reporting Trend in Indian Companies as per GRI Framework: A Comparative Study. *South Asian Journal of Business and Management Cases*, 4, pp. 62–73. doi:10.1177/2277977915574040.

Goel, P. and Misra, R. (2017). Sustainability Reporting in India: Exploring Sectoral Differences and Linkages with Financial Performance. *Vision: The Journal of Business Perspective*, 21(2).

Gray, R. (1992). Accounting and Environmentalism. An Exploration of the Challenge of Gently Accounting for Accountability, Transparency and Sustainability. *Accounting, Organizations and Society*, 17(5), July, pp 399–425.

Hacking, T. and Guthrie, P. (2008). A Framework for Clarifying the Meaning of Triple Bottom-Line, Integrated, and Sustainability Assessment. *Environmental Impact Assessment Review*, 28, pp 73–89.

Henri, J-F. and Journeault, M. (2010). Eco-Control: The Influence of Management Control Systems on Environmental and Economic Performance. *Accounting, Organizations and Society*, 35(1), pp 63–80.

Jones, P., Hillier, D. and Comfort, D. (2017). The Sustainable Development Goals and the Tourism and Hospitality Industry. *Athens Journal of Tourism*, 4(1), pp 7–18.

Joseph, K., Eslamian, S., Ostad-Ali-Askari, K., Nekooei, M., Talebmorad, H. and Hasantabar-Amiri, A. (2019). Environmental Impact Assessment as a Tool for Sustainable Development. In: Leal Filho, W. (eds.), *Encyclopedia of Sustainability in Higher Education*. Cham. Springer, p 19.

Kaplan, R.S. and Norton, D.P. (1992). The Balanced Scorecard: Measures That Drive Performance. *Harvard Business Review*, January–February, pp 71–79.

Kaplan, R.S. and Norton, D.P. (2004a). *Strategy Maps: Converting Intangible Assets into Tangible Outcomes*. Boston. HBS Press.

Kaplan, R.S. and Norton, D.P. (2004b). Measuring the Strategic Readiness of Intangible Assets. *Harvard Business Review*, 21, February, pp 52–63.

Kates, R.W., et al. (2001). Sustainability Science. *Science*, 292, pp 641–642.

Kolk, A. (2004). A Decade of Sustainability Reporting: Developments and Significance. *International Journal of Environment and Sustainable Development*, 3(1), pp 51–64.

KPMG. (2015). Currents of Change: KPMG Survey of Corporate Responsibility Reporting 2015. https://assets.kpmg.com/content/dam/kpmg/pdf/2016/02/

kpmg-international-survey-of-corporate-responsibility-reporting-2015.pdf. Accessed 28 December 2017.

Kumar, K. and Prakash, A. (2019). Examination of Sustainability Reporting Practices in Indian Banking Sector. *Asian Journal of Sustainability and Social Responsibility*, 4(2). https://doi.org/10.1186/s41180-018-0022-2.

Mathews, M.R. (1997). Twenty-Five Years of Social and Environmental Accounting Research: Is There a Silver Jubilee to Celebrate? *Accounting Auditing & Accountability Journal*, 10(2), pp 481–531.

Ness, B., Urbel-Piirsalu, E., Anderberg, S. and Olsson, L. (2007). Categorising Tools for Sustainability Assessment. *Ecological Economics*, 60, pp 498–508.

Pattanaro, G. and Donato, S. (2018). Sustainable Development Goals: A Business Opportunity for Tourism Companies? *Economic Problems of Tourism*, 43, pp 21–27. doi:10.18276/ept.2018.3.43-02.

Pedersen, C. (2018). The UN Sustainable Development Goals (SDGs) Are a Great Gift to Business! *Procedia CIRP*, 69, pp 21–24. doi:10.1016/j.procir.2018.01.003.

Pillai, K.V., Slutsky, P., Wolf, K., Duthler, G. and Stever, I. (2017). Companies' Accountability in Sustainability: A Comparative Analysis of SDGs in Five Countries. In: Servaes, J. (ed.), *Sustainable Development Goals in the Asian Context*. Singapore. Springer/Urbana-Champaign, IL. Research in Accounting, pp 85–116.

Schaltegger, S. and Burritt, R.L. (2000). *Contemporary Environmental Accounting: Issues, Concepts and Practice*. Sheffield. Greenleaf Publishing.

Schaltegger, S. and Sturm, A. (1992). *Environmentally Oriented Decisions in Firms: Ecological Accounting Instead of LCA: Necessity, Criteria, Concepts*. Bern/Stuttgart. Haupt.

Schaltegger, S. and Wagner, M. (2006). Integrative Management of Sustainability Performance, Measurement and Reporting. *International Journal of Accounting, Auditing and Performance Evaluation*, 3(1), pp. 1–19.

Scheyvens, R., Banks, G. and Hughes, E. (2016). The Private Sector and the SDGs: The Need to Move Beyond 'Business as Usual'. *Sustainable Development*, 6(24), pp 371–382.

Schoenmaker, D. and Schramade, W. (2019). Investing for Long-Term Value Creation. *Journal of Sustainable Finance & Investment*, pp. 1–22. doi:10.1080/20430795.2019.1625012.

SDG Fund. (2016). Sustainable Development Goals Fund. Universality and the SDGs: A Business Perspective. http://www.sdgfund.org/sites/default/files/Report-Universality-and-the-SDGs.pdf.

Sharma, A. (n.d.). Sustainability Reporting Trends in India. Available at http://www.iodonline.com/Articles/Arvind%20Sharma%20%20Sustainability%20Reporting%20Trends%20in%20India_KPMG.pdf.

Sinha, R. and Datta, M. (2019). Institutional Investments and Responsible Investing. In: Ziolo, M. and Sergi, B. (eds.), *Financing Sustainable Development*. London. Palgrave Studies in Impact Finance.

Stechemesser, K. and Guenther, E. (2012). Carbon Accounting: A Systematic Literature Review. *Journal of Cleaner Production*, 36, pp 17–38. doi:10.1016/j.jclepro.2012.02.021.

Venning, J. and Higgins, J. (2001). *Towards Sustainability: Emerging Systems for Informing Sustainable Development*. Sydney. UNSW Press.

Verma, B. (2002). *Corporate Social Accounting and Disclosure Practices in Public Undertaking*, 1st ed. Jaipur. RBSA Publishers.

Vitale, G., Cupertino, S., Rinaldi, L. and Riccaboni, A (2019). Integrated Management Approach Towards Sustainability: An Egyptian Business Case Study. *Sustainability*, 11, p 1244.

Waas, T., Hugé, J., Block, T., Wright, T., Benitez-Capistros, F. and Verbruggen, A. (2014). Sustainability Assessment and Indicators: Tools in a Decision-Making Strategy for Sustainable Development. *Sustainability*, 6, pp 5512–5534. doi:10.3390/su6095512.

Wells, M.C. (1978). Accounting for Common Costs, Center for International Education and Accounting for Accountability, Transparency and Sustainability. *Accounting, Organizations and Society*, 17(5), pp 399–425.

Wulf, C., Werker, J., Zapp, P., Schreiber, A., Schlör, H. and Kuckshinrichs, W. (2018). Sustainable Development Goals as a Guideline for Indicator Selection in Life Cycle Sustainability Assessment. *Procedia CIRP*, 65. doi:10.1016/j.procir.2017.11.144.

Chapter 14

Sustainable financial reporting in the context of ensuring sustainability of financial systems

Beata Zofia Filipiak and Marek Dylewski

Introduction

The Rio Summit, held in 1991, established that accounting should play a crucial role in the global sustainable development agenda. Corporate sustainability has been presented as the ultimate goal for corporations, meeting the needs of the present without compromising the ability of future generations to meet their own needs (Thomsen, 2013). Reporting information regarding sustainable development became a means by which major companies could show their concern for increasing the transparency of the activities they conduct and for promoting corporate responsibility (Thoradeniya et al., 2015; Adams & Whelan, 2009).

It is assumed that corporate reporting is the responsibility of an organisation to indicate the impact of an organisation's decisions and activities on society and the environment by means of transparent and ethical behaviour (Blowfield & Murray, 2011; Jonikas, 2013). Sustainability reporting is a tool that can be used to increase transparency and accountability in the issues that traditional financial reporting does not deal with. Companies respond to societal pressures using sustainable reporting as a tool for confirming the socially responsible behaviour mandated by the external environment in which they conduct their activity (Manes-Rossi et al., 2018). In accordance with the voluntary disclosure theory, companies also perceive the instrumental role of presenting information pertaining to social responsibility in improving economic performance (Clarkson et al., 2008; Oh et al., 2017; Clarkson et al., 2013; Dhaliwal et al., 2011) and submit such data to improve their image and reduce the negative effect of their own activities (Carp et al., 2019).

Sustainability reporting is the reporting by institutions or organisations on the economic, environmental, and social impacts (ESG factors) of their daily activities. An increasing number of companies, banks and institutions are implementing sustainability reporting. The objective of implementing sustainability reporting may be different, as shown by the research:

- of organisational legitimacy (Hedberg & Malmborg, 2003),
- account the oversight of the board in order that it is exercised,

- the arrangement of sustainability responsibilities Kolk, 2004),
- attention to compliance, ethics and external verification (Kolk, 2004),
- reporting stakeholder issues and achievements in engaging stakeholders (Mulkhan, 2013).

Studies have found that sustainability reporting has a positive effect on the stock prices of real estate companies. Ansari et al. (2015), Loh et al. (2017) and Lourenço et al. (2014) have shown the usefulness of sustainability reporting. In addition, studies show that sustainability reporting has an impact on the value relevance of the financial statements, which include earnings per share (EPS), earnings per share change (EPSC) and book value per share (BVPS) (Sutopo et al., 2018). However, previous research findings also show that there are weaknesses in sustainability reporting (Hedberg & Malmborg, 2003; Farneti & Guthrie, 2009) and that sustainability reporting provides more qualitative information on financial value than quantitative information (Lins et al., 2008; Hristov et al., 2019).

An important issue after the financial crisis, and also in the context of taking environmental issues into account, is the stability of financial systems. It has been increasingly accepted by monetary and supervisory authorities and financial institutions that climate- and sustainability-related factors are a source of financial risk and fall within the financial stability mandates of central banks and supervisors (NGFS, 2018).

In striving for stability, financial institutions accept non-financial reporting and go further by themselves to prepare non-financial reports. Bearing in mind the threats to the financial system resulting from climate and environmental issues, they also recognise that other ESG factors are important to the financial system (Benlemlih & Bitar, 2018; Friede et al., 2015; In et al., 2017; Pereira da Silva, 2019).

One important question these days is whether, and how, financial reporting can affect the European Union's (EU's) objective of sustainable development and ensure the sustainability of financial systems. Financial reporting and accounting has proved over time to be a powerful practice, which is embedded in an institutional context and shapes economic and social processes (e.g. Soll, 2015; Palea, 2018). An important aspect is the use of sustainable financial reporting to ensure the sustainability of financial systems.

The objectives of this chapter are firstly, a review of non-financial reporting, followed by a presentation of its historical development. Finally, this chapter will provide an overview of new trends and challenges in non-financial reporting.

Sustainable financial reporting: evolution paths and new trends

Sustainability reporting is an issue that has been the focus of the business world and society oriented to the respect of resources in recent years

(Tvaronavičienė et al., 2017; Sutopo et al., 2018). Sustainability reporting is the reporting by companies, institutions or organisations on the economic, environmental and social impacts caused by their daily activities. The scope of financial reporting includes the financial statements and disclosures outside of the financial statements that are the products of accounting (Sutopo et al., 2018). A firm's survival can no longer be traced back only to an economic dimension of profit maximisation, but rather has to be included in a broader discourse encompassing the way in which the firm manages the risks arising from the social and environmental impacts of its activities in the medium and long term and demonstrates them to be socially responsible (Hackston & Milne, 1996; Milne & Gray, 2007; Manes-Rossi et al., 2018). Stakeholders and report makers have become increasingly focused on the inclusion of non-financial information in annual reports. Non-financial information refers to a broad range of themes and issues such as environmental and social policies and impacts (e.g. resource and energy use, greenhouse gas emissions, pollution, biodiversity, climate change, waste treatment, the health and safety of employees, gender equality and education) and is pivotal in improving accountability and transparency towards stakeholders (Kolk, 2004; Prado-Lorenzo et al., 2009; Bebbington et al., 2014, Manes-Rossi et al., 2018).

Sustainability reports allow companies to demonstrate that they are socially responsible and are a powerful tool for improving communication with stakeholder groups and society by enhancing the transparency and accountability of non-financial information and strengthening social responsibility (Patten & Zhao, 2014). They are often called non-financial reports or non-financial information disclosure reports. This approach has two sources, one of which is the presentation of financial data in sustainability reports. The second source is related to the fact that financial reports contain data sourced from accounting that presents financial data.

The issue of qualitative characteristics of non-financial information contained in annual reports is regulated, on the one hand by the International Financial Reporting Standards (IFRS), but also results from the habits and the desire to build the credibility and responsibility of organisations and institutions. This means that from the IFRS point of view, these are the properties of the report that make the information it contains useful for users. They include intelligibility, relevance, reliability and comparability. On the other hand, such features as relevance, faithful representation, comparability, verifiability, timeliness and understandability are pointed out (IFRS, 2019).

Investment risk assessment criteria regarding corporate social responsibility (CSR), ESG and responsible business conduct (RBC) are starting to play an increasingly important role in the strategies of many institutions and organisations in mature Western European markets and in the United States. Environmental awareness and the pursuit of stability are also growing among institutions and organisations. This stability is required differently, but stability is indicated as a function of sustainability. Heal (1998) suggests that the essence of sustainability is defined by the following three axioms:

- the treatment of the present and the future that places a positive value on the very long term,
- the recognition of all the ways in which environmental assets contribute to economic wellbeing,
- the recognition of the constraints implied by the dynamics of environmental assets.

This approach, which has been under development for years now, indicates that organisations and institutions, in striving for stability, must maintain the rules of sustainability (Mensah, 2019; Ziolo et al., 2019). It is also indicated that stakeholders in mature capital markets are convinced that the assessment and valuation of specific categories of business risks related to their business will be transparently communicated through non-financial (ESG) data. Thus, it will be easier to receive and be more precise.

Financial reporting is conducive to building trust, as well as perhaps contributing to the possibility of an expanded dialogue with stakeholders and improving processes and systems, as processes can be examined and improved thanks to accounting and internal control management and decision-making, which can lead to cost reduction (Joshi & Li, 2016). Non-financial reporting is not only conducive to communicating vision and strategy but also allows for an assessment of the progress in the implementation of the vision and strategy through the necessary analysis of strengths and weaknesses and the involvement of stakeholders in the processes of sustainable development reporting (Boubakary & Moskolaï, 2016; Niño-Muñoz et al., 2019). Non-financial reporting also reduces the cost of violations, as measuring sustainability can help one avoid costly violations. These activities lead to the strengthening of the stability of organisations and institutions, thus enabling them to achieve a competitive advantage.

The relationship between negative externalities and sustainable development and ensuring the sustainability of financial systems is not visible at the general level of definitions, but the subcategories of sustainable finance are strictly related to pollution, smog, noise and social exclusion which are found in definitions of environmental finance, green finance, carbon finance, development finance, responsible finance and financial stability factors. The studies highlighting the multi-sectoral range of impact pay special attention to ESG risk and the role of public funds (income and expenditure instruments).Wang & Zhi, 2016; Mazzucato & Semieniuk, 2018; Falcone, 2018; Alińska et al., 2018; Falcone et al., 2018; Falcone & Sica, 2019; Ziolo et al., 2019).

The development of the concept of sustainable development is the only right way to ensure further socio-economic development and the adoption of its principles for implementation at both macroeconomic and microeconomic levels related to:

- the growing degradation of the natural environment, which affects both the creation of ecological barriers to humans doing business and threatens their existence on Earth,
- emerging demographic and social problems on a global scale, related to the growing population, an increase in poverty and unemployment and an increase in social inequalities,
- increasing the awareness of societies about emerging ecological threats that contribute to a reduction in their quality of life and inhibit the further development of civilisation.

These factors have forced the real economy sector, financial institutions and the public sector to change their perception of information. Considering, in brief, that accounting provides information for decision-making purposes, it has become necessary to modify accounting principles and systems. There is a need for sustainability accounting and sustainability management control (Lamberto, 2005). Together with the idea of sustainability accounting, sustainability reporting is developing naturally. There are many reasons for adopting sustainability-oriented reporting, ranging from external pressures from local communities, media and consumers or coming from the responsiveness of management.

In world literature, standards, guidelines and other substantive documents, we encounter a lack of explicitness in terms defining the concept of the reporting of non-financial data. In literature referring to non-financial reporting, the following terms are used: "Environment & Social & Governance", "Sustainability Report", "Corporate Responsibility Report", "Corporate Citizen Report", "Sustainable Development Report", "Environment & Health & Safety Report", "Environment & Social", "Environment & Health & Safety & Community", "Corporate Social Responsibility", "Social and Community" and "Non-financial Data Report". In the accounting nomenclature, as well as in the solutions of many countries, terms such as the social report, ESG data report, sustainable development report and CSR report have been adopted. The emergence and development of sustainable financial reporting has taken three paths: a path closely linked to the directions of sustainable development; a path of business self-regulation related to information needs and social responsibility; and a path of civic pressure to include non-financial factors in financial statements.

Studies of the path closely linked to the directions of sustainable development include research on factors and performance indicators (see Figure 14.1). Actions on these paths intertwine and often lead along a common path, i.e. a path closely linked to the directions of sustainable development and a path of business self-regulation related to information and social

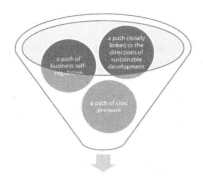

sustainable financial reporting

Figure 14.1 Three paths of sustainable financial reporting
Source: Authors' own elaboration.

needs. They interpenetrate using research, findings (designations), good practices and legal regulations.

The path closely linked to the directions of sustainable development was developed in several directions. The first group includes strategic planning and managerial systems that – starting from the strategic guidelines – identify the goals and the roadmap to follow at various organisational levels. This group is an important direction of non-financial disclosures and reporting (Truant et al., 2017; Breuer et al., 2019). As part of the work welcoming the suggestions of several authors, such planning and control systems have begun to account for sustainability goals and related risk indicators (Dias-Sardinha & Reijnders, 2001; Figge et al., 2002; Schaltegger & Wagner, 2006; Henri & Journeault, 2010; Bouten & Hoozée, 2013; Epstein & Buhovac, 2014; Truant et al., 2017). The researchers also discussed a theorisation of management control systems in addressing sustainability issues, creating the so-called model of a sustainability control system (Gond et al., 2012). The literature review presents the notion of "eco-control" as the application of financial, strategic control methods to environmental management (Henri & Journeault, 2010); the role of planning a sustainable strategy (Bonacchi & Rinaldi, 2007); the introduction of specific environmental budgeting techniques (Roth, 2008); and case studies on environmental/material cost accounting systems and environmental investment appraisal (Herzig et al., 2012). Despite tools such as the balanced scorecard, internal performance measurement, reporting and management control mechanisms integrating financial and non-financial strategic measures (Kaplan & Norton, 1996; Möller & Schaltegger, 2005; Journeault, 2016; Truant et al.,

2017), few studies have focused on their application to sustainability management (Schaltegger & Wagner, 2006; Hubbard, 2009). The importance of the balanced scorecard (SBC), albeit oriented towards sustainable development, is also pointed out (Figge et al., 2002; Rafiq et al., 2020).

The second group refers to external non-financial communication and reporting tools. Turning now to the second group (external reporting), several social and environmental reporting frameworks were developed in the last part of the 20th century. There are many reasons for adopting sustainability-oriented reporting, ranging from external pressures from local communities, media and consumers, to those which arise from the responsiveness of management. Another motivation could be the search for greater workplace legitimacy (O'Dwyer, 2002; Cho & Patten, 2007; Cantele, 2014). There are different traditions and customs depending on the country and continent (see: Astupan & Schönbohm, 2012; Crane, 2018; Farneti & Guthrie, 2009; Corporate . . ., 2019) This situation normalises and feeds into the framework of the second path – the path of business self-regulation.

Turning now to **the second path** (a path of business self-regulation related to information needs and social responsibility), two directions of action can be set here. The first is related to accounting, its development and support for scientific research on the use of accounting information in non-financial reporting (the application of scientific research in accounting). The second direction is related to the activities of the institutional sense, often normative, aimed at formulating guidelines.

As part of the direction related to accounting, several social and environmental reporting frameworks were developed at the end of the 20th century. Organisations can choose to disclose selected information about their social and environmental impacts, as well as their policies, in separate stand-alone reports or as part of their annual reports by managing interactions between the organisation or institution and the external environment – stakeholders (Unerman, 2000; De Villiers et al., 2014).

Gray (1993) is credited with much of the conceptual development of sustainability accounting. He identifies three different methods of sustainability accounting: 1) sustainable cost, 2) natural capital inventory accounting and 3) input–output analysis (Lamberto, 2005). Gray's achievements are important in terms of quality and targeting non-financial reporting. The practical problems of valuing external costs such as pollution have been well defined (Mathews, 1993; Pearce & Turner, 1990). Any damage to critical natural capital would, in theory, be valued at infinite cost because it is irreplaceable, leading to the conclusion that the activities of an organisation which damage critical natural capital are unsustainable (Gray, 1994). Research indicates that the science of ecology does not provide clear and unchallenged solutions to environmental problems (Holland & Petersen, 1995; Bedenik &

Barišić, 2019), and whilst placing costs on a range of possible solutions to environmental problems, may prove exhausting (Mathews, 1995).

The achievements of Kreuze and Newell (1994) are important for non-financial reporting. They pointed out that by identifying the environmental costs and tracing them to their sources, companies can produce relevant information for engineers, production personnel and marketing staff to help keep their products competitive. Hammer and Stinson (1995) presented the importance of measuring environmental costs and accounting for them as an intrinsic part of corporate decision-making by assigning the environmental costs to products or processes. The advent of activity-based costing and life cycle costing enables the identification and measurement of specific environmental costs (Kreuze & Newell, 1994; Ditz et al., 1995).

Institutional theorists also recognise that institutional expectations do not apply uniformly to all organisations, and firms can respond differentially to these isomorphic pressures by adopting structural elements selectively or ceremonially by decoupling from operational decision-making (Meyer & Rowan, 1977). Firms conform to institutional pressures by incorporating elements of sustainability to financial reporting that are legitimised externally, and conforming organisations are rewarded with social stability, reduced uncertainty, extra resources, access to stable sources of financing and survival capabilities (Meyer & Rowan, 1977; North, 1990; Scott, 2001, 2010). A common understanding of the basic elements of financial reporting is needed to compare them. Thus, it is necessary to properly account for them in terms of accounting and standardisation (Lamberto, 2005; Burritt & Schaltegger, 2010).

Bebbington and Thomson (2007) discussed the role of social and environmental accounting and reporting (SEAR) studies by citing the works of Beck (Beck, 2006; Beck & Holzer, 2007). The literature review stressed (Bedenik & Barišić, 2019; Ziolo et al., 2019) the importance of ecological risks and hazards, with the need to introduce a risk culture of social environmental accounting because of the ability of accountants to fully capture their dimension in time and space, measurement, evaluation and calculation. The discourses and practices of environmental accounting and reporting are an attempt to respond to these perceived weaknesses to risk identification and disclosure by making impacts that are currently ignored or hidden by mainstream accounting visible (Bebbington & Larrinaga, 2014).

Practicing sustainability also creates opportunities for new product markets, business model innovations and value creation. Thus, sustainability management essentially becomes an integral part of the overall business strategy. Simple management accounting tools (which are being developed) are used, and advanced instruments, procedures and systems based on the use of advanced accounting in strategic terms are created. New business strategy based on sustainable development requires specific information, key performance indicators and decision-making criteria.

In order to provide relevant information for the business strategy based on sustainable development, it may be necessary to use appropriate tools, such as the sustainability balanced scorecard and strategy maps (Kaplan & Norton, 1996; Figge et al., 2002; Möller & Schaltegger, 2005), eco-control (Henri & Journeault, 2010; Schaltegger & Sturm, 1995), or sustainability management control (Burritt & Schaltegger, 2010). Environmental cost information, unified and ready to use, can influence key business decisions such as product costing and pricing, the product mix, regulatory negotiations, risk management, product design and differentiation, labelling and tax planning (Joshi & Krishnan, 2010). This is a serious premise for uniform recognition of performance indicators of sustainability in a non-financial statement.

Along with scientific research on the concept of non-financial reporting, the work of the institution is carried out (this pertains to the second direction related to activities in the institutional sense). A number of organisations are developing sustainability reporting standards for firms, with the goal of making them useful to external stakeholders. An article by Hales et al. (2016) compares and contrasts the key features of the reporting models of four major institutions in non-financial reporting, namely the Global Reporting Initiative (GRI), the International Integrated Reporting Council (IIRC), the Sustainability Accounting Standards Board (SASB), and the Carbon Disclosure Project (CDP). The proposed reporting standards and models differ in terms of target stakeholder groups, the definition of materiality, data collection, aggregation agency, the scope of performance metrics and report generation models.

Numerous studies pay special attention to projects aimed at integrating the reporting of non-financial data with financial data in a common report (Arvidsson, 2011). Widespread non-financial reports covering key areas, include:

- GC Communication on Progress – CoP,
- AA1000,
- the evolution of ISO standards,
- OECD guidelines for multinational enterprises,
- GRI guidelines,
- Integrated reports.

The increase in interest in non-financial information has resulted in the emergence of a new, rapidly developing stream, namely integrated reporting. The Global Compact idea is expressed in 10 basic principles relating to human rights, labour standards, the environment and anti-corruption practices, resulting from the Universal Declaration of Human Rights, the charter of basic principles of International Labour Organisation law and the findings of the Earth Summit in Rio de Janeiro. The basic report is the

GC Communication on Progress (CoP) (Basic Guide . . ., 2019). In relation to human rights, issues such as compliance with and support for the protection of internationally recognised human rights and the elimination of all human rights violations by the company are reported. As part of labour standards, reporting includes freedom of association and recognition of the right to collective negotiations; the elimination of slave labour and forced labour, including child labour; and prevention of discrimination in the field of employment. In terms of environmental protection standards, reports include a preventive approach to environmental problems, initiatives promoting greater environmental responsibility and actions for the development and dissemination of environmentally friendly technologies (WBCSD, 2004).

Anti-corruption is an important and well-developed standard. Reporting under this standard includes counteracting corruption in all its forms, including bribery and extortion (CoP Policy, 2015). It should be emphasised that this initiative is expanding (new activities and initiatives are emerging) and includes a number of factors which are important for sustainable development that capture ESG risks in detail. It should be noted that other external factors are also important. These include but are not limited to fraud, money laundering and public ethnocentrism, which have the same effects as corruption. On the one hand, it is difficult to expect that the reports will have room for disclosure of the factors identified; but on the other hand, it is important to indicate in non-financial reports how companies, banks or institutions protect themselves or what actions they take in this regard. While sustainability is largely associated with do-gooders, non-financial reports discuss whether and how fraud might also be an issue in sustainability departments (Dunn, 2014). The preparation of non-financial reports is exposed to fraud because sustainability managers face mounting pressure and have opportunities to manipulate reports due to an immature control environment. Not only control but also morality standards become important and underlining in non-financial reports to light the importance of a clear commitment from executives to sustainability to prevent sustainability fraud. Showing morality standards in non-financial reports is also aimed at assuring that the negative impact on reporting caused by external stakeholders is eliminated, specifically the bonus of sustainability index scores (Steinmeier, 2016). In the literature on the subject, research has shown that the inclusion of non-financial information on past occurrences to fraud, money laundering and public ethnocentrism can create new business opportunities. It is one of the possibilities of impact eliminating these phenomena and obtaining assurances (Eremina et al., 2020). Following the example of financial institutions (e.g. mBank, 2020), one can apply a "zero tolerance" principle to corruption, money laundering and public ethnocentrism in all its forms and include information in non-financial reports that institutions, enterprises or public entities have acted on so as to eliminate these factors.

It should be remembered that companies willing to implement the idea of the Global Compact can declare adherence to the Minimum Standard (a set of tools recognised as a set of basic solutions constituting the starting point for creating a non-financial report) and the Aspiration Standard (an open set containing recommended practices, and in particular the development of the Minimum Standard). Therefore, stakeholders should accept with high praise actions for extending standards with information related to the impact of external factors such as fraud, money laundering and public ethnocentrism. They should also positively accept and react negatively to any lack of information – that there is "zero tolerance" for the actions of managers related to the preparation of reports and concealing facts.

The AA1000 standards were developed by the Institute of Social and Ethical AccountAbility (AccountAbility) in Great Britain and were first presented in November 1999. The AA1000 standard is targeted at stakeholders of organisations and institutions. The purpose of this standard is to support management processes and communication with stakeholders. The AA1000 series consists of the following standards (Leipziger, 2017; AccountAbility 1000 . . ., 1999):

- AA1000 Principles of Responsibility (AA1000APS),
- AA1000 Verification (Accountability Assurance Standard – AA1000AS),
- AA1000 Stakeholder Engagement Standard (AAA1000SES).

This standard allows for the assessment of data compliance with the facts and the quality of reported information on the organisation's activities, in particular in the area of management and results.

The beginnings of financial reporting are rooted in life cycle assessment (LCA), which began in the 1970s. LCA tracks the impact of business activity from the extraction of raw materials through manufacturing, use and final disposition. In LCA, factors are tracked and impacts are classified into resource depletion, human health and environmental health. Human well-being is not analysed beyond health issues. LCA also affected the conceptual development of the ISO 14000 family of environmental management system guidelines (Tibor & Feldman 1996). These activities led to the development of formalised environmental management systems (EMSs) (Vann & White, 2004).

The next breakthrough was the implementation of EMSs, which were created to help companies manage their environmental costs. They were established not only to comply with environmental regulations but also to help companies reduce and prevent environmental costs.

The demand for additional indicators about corporate economic, environmental and social performance (hence the term triple bottom line) has been driven by stockholder initiatives supported by the Coalition for Environmentally Responsible Economies (CERES, 2004), by stock indices such as FTSE4GOOD (2004) and the Dow Jones Sustainability Indexes (DJSI, 2004) and by the requirements of ISO 14001 (ISO, 2004). The International Organization for Standardization (ISO) implements the ISO 14000 standard, the ISO, 2004 standard and environmental certificates (ISO, no date).

The international ISO 26000 Standard – Guidance on Social Responsibility was issued in 2010. ISO 26000 guidelines emphasise the value of publishing the organisation's results in the field of CSR for internal and external stakeholders. Non-financial reporting issues are described in seven key areas (ISO, 26000):

- Organisational order,
- Human rights,
- Relations with employees,
- The environment,
- Fair market practices,
- Issues related to clients/consumers,
- Social commitment and development.

The ISO 26000 standard does not offer guidance on non-financial reporting; the content of this standard covers a very similar thematic scope as the GRI reporting guidelines.

The OECD guidelines for multinational enterprises were first developed in 1976. Their purpose was to ensure that international corporations comply with the policies of the countries in which they operate. The guidelines were aimed at defining voluntarily applied principles and standards of responsible behaviour of organisations and institutions. In 2000, these guidelines were updated, and they were further modified in 2011. The third chapter of the OECD document presents issues of disclosure of information by institutions and organisations, in particular in two areas. The first area, a set of recommendations, concerns the timely and reliable disclosure of relevant information about the organisation or institution, including its financial position, results, ownership and supervision. Sufficient disclosure of the remuneration of the management board and CEO is also expected. The second set of guidelines addresses issues regarding society, the environment and risk (OECD Guidelines . . . 2011).

In 1997, in partnership with the United Nations Environment Programme, CERES started the GRI (Global Reporting Initiative, 2002, p. 1). The Sustainability Reporting Guidelines (GRI 1 – G1) were published in 2000 and revised in 2002 (GRI 2 – G2). They recommend measures within all three domains of sustainability. The University of Florida Sustainability Indicators August 2001 were "Published in Accord with The Global Reporting Initiative Sustainability Reporting Guidelines, June 2000" (Chesnes and Newport, 2000, front cover). Major accounting firms also recognise the problem and the need to strengthen their efforts (Deloitte, Touche, Tohm-atsu . . ., 2002).

The economic category of indicators is designed to supplement financial information contained in conventional financial accounting reports, providing information concerning the impact of an organisation's activities on 1) the economic circumstances of stakeholders and 2) local, national and global economies (GRI, 2002). In the period of 2006–2013, subsequent versions of the guidelines came into force – G3, G3.1 and G4, respectively. In November 2016, the Global Sustainability Standards Board (GSSB) announced the latest version of the guidelines – the so-called GRI Standards. The development of the global reporting framework published by GRI in chronological order is shown in Figure 14.2. Table 14.1 presents the set of aspects in terms of ESG factors in the GRI guidelines and standards. It is worth pointing out that the organisation may use all or some of the standards, or even selected parts of them. The last part of GRI Standards is a glossary. The modular structure of GRI standards is presented in Figure 14.3.

The GRI standards has a changed, modular structure and set of indicators. There are general standards, applicable to any organisation and institution preparing a report on sustainable development (3 standards) and thematic standards: economic (6 standards), environmental (8 standards) and social (19 standards).

Recapitulating, GRI reporting standards are the best-known and most commonly used document that helps prepare non-financial reports for organisations around the world. The popularity of GRI regulations is primarily due to two reasons. First of all, while creating standards, GRI cooperates with many international organisations and creates a kind of network of stakeholders and a circle of experts from various environments. Secondly, the undoubted advantage of reporting according to GRI guidelines/standards is their flexibility.

Figure 14.2 Changes in the GRI guidelines and standards

Source: Authors' elaboration.

Table 14.1 Set of aspects in terms of ESG factors in the GRI guidelines and standards

Factors	Category	Aspects
Economic	Direct economic impacts	• Customers • Suppliers • Employees • Providers of capital • Public sector
Environmental	Environmental	• Materials • Energy • Water • Biodiversity • Emissions, effluents, and waste • Suppliers products and services • Compliance • Transport • Overall
Social	Labour practices and decent work	• Employment • Labour/management relations • Health and safetyTraining and education • Diversity and opportunity
	Human rights	• Strategy and management • Non-discrimination • Freedom of association and collective bargaining • Child labour • Forced and compulsory labour • Disciplinary practices • Security practices • Indigenous right
	Society	• Community • Bribery and corruption • Political contributions • Competition and pricing
	Product responsibility	• Customer health and safety • Products and services • Advertising • Respect for privacy

Source: Authors' elaboration on GRI 2 (2002) and Lamberto (2005).

The aforementioned integrated reporting is a new trend emerging in the area of reporting. The key to supporting integrated reporting at the international and national level is the content of paragraph 47 of the final document of the Rio +20 UN Summit.

Integrated reporting combines relevant information on the organisation's strategy, management, results and future prospects in a way that reflects

Modules	Module 1 Universal standards (GRI 100)	Module 2 Thematic standards (GRI 200-400)	Module 3 Glossar
Scope	GRI 101 standard: Basic information	GRI 200: Economic standards	-
	GRI 102 standard: Profile indicators	GRI 300: Environmental standards	-
	GRI 103 standard: Management approach	GRI 400: Social Standards	-

Figure 14.3 The modular structure of GRI standards

Source: Authors' elaboration based on www.globalreporting.org/standards/gri-standards-download-center/.

the economic, social and environmental context in which it operates. Such reporting is a clear and concise representation of how an organisation demonstrates its responsibility and how it creates value now and in the future.

Integrated reporting combines the most important elements of information currently presented in separate documents (the financial report, the management board report, the corporate governance statement and the sustainable development report) into a coherent whole. As for the question: "What is an integrated report?", the simplest definition found is that it is a single document that contains a company's financial and non-financial ESG performance (Eccles & Saltzman, 2011; Wensen et al., 2011).

The integrated report explains how non-financial data affect an organisation's ability to create and sustain value in the short, medium and long term (A4S, 2010, 2011; Flower, 2015; Stacchezzini et al., 2016). In the coming years, the significance of sustainable development reports and integrated reports are expected to decline, as shown by the first public consultation report of the new generation of GRI G4 guidelines (A4S, GRI 2010). GRI G4 (G4, Sustainability Reporting) reporting is becoming one of the basic trends related to non-financial reporting. It is noteworthy that the integrated reporting was a separate framework, building on the IFRS and the GRI (IIRC, 2017).

The combination of accounting, advanced methods of management accounting and effects of research that are the result of sustainable development is just one example of the interpenetration of paths and the combination of a path closely linked to the directions of sustainable development, a path of business self-regulation related to information requirements and social responsibility.

The third path is very different in different countries, not only in terms of continents, but above all in particular countries. Activities on this path result primarily from the level of social capital, the level of social trust, social involvement and strong social movements, bringing together citizens' activity in the area of sustainability, solidarity and social activity controlling the commercial sector and forcing social responsibility. Consumer boycotts or other social campaigns, organised through an increasingly popular network and social media, translate (more or less – unfortunately sometimes less) into actual consumer behaviour (Young et al., 2010; Lavorata, 2014; Seegebarth et al., 2016).

The scope and presentation of sustainable development issues in the non-financial statement[1]

Civic pressure, the emergence of the need to publish reports, social responsibility and other factors caused by discussion in the environment, in literature and at the level of governments and organisations, have resulted in the emergence of the position of the EU. Following changes in the scope of the requirements of shaping sustainable development and ensuring stability, the EU has adopted Directive 2014/95/EU, which pointed to the concept of non-financial reporting.

Sustainability reporting is about measuring, disclosing and being responsible to internal and external stakeholders for results and efficiency in addressing these issues. The sustainability report should provide balanced and relevant information on the performance of the organisation (institution, enterprise) in addressing sustainability issues, including both positive and negative aspects. Such reports can be used for the following purposes, among others:

- setting a benchmark and basis for assessing the performance of an organisation in addressing sustainable development issues with respect to law, norms, codes, standards and voluntary initiatives,
- demonstrating the impact of the organisation and being influenced by sustainability expectations,
- comparing the performance of an organisation over time and against other organisations.

In 2014, Directive 2014/95/EU on the disclosure of non-financial and diversity information was issued. On 26 June 2017, the European Commission published guidelines on reporting in the field of non-financial information. In the commission's opinion, these guidelines may be helpful for companies to disclose material non-financial information in a consistent and comparable manner.

On 20 June 2019, the European Commission published guidelines on reporting climate-related information, which are a supplement to the existing guidelines on non-financial reporting that are still in force. In particular, the EU has the following legal provisions governing non-financial reporting:

- Directive 2014/95/EU of the European Parliament and of the Council of 22 October 2014 amending Directive 2013/34/EU with regard to the disclosure of non-financial and diversity information by certain large undertakings and groups,
- Communication from the Commission – Guidelines on non-financial reporting (methodology for reporting non-financial information) (2017/C 215/01),
- Communication from the Commission – Guidelines on non-financial reporting: Supplement on reporting climate-related information (2019/C 209/01),
- legal regulations in force in EU member states.

The European Commission committed to reviewing the Non-Financial Reporting Directive in 2020 as part of its strategy to strengthen the foundations for sustainable development and stability.

General reporting standards and guidelines of non-financial reporting are loaded with imperfections due to their limited ability to take into account the specifics of a given industry and sector, as well as the specifics of the country in which an institution or organisation operates. Despite the flexibility of report preparation and national and industry specificity, non-financial reports should meet specific requirements. These requirements can be included in the framework of reporting rules, which – after structuring – will allow organisations to answer two questions: how to report and what to report (see Figure 14.4).

The following rules (principles), which result from balanced accounting and the provisions of the EU directive, are of key importance in the preparation of the non-financial report:

- materiality – all policies, procedures and events that significantly affect the business and the situation of the enterprise or institution and its economic, social and environmental results should be indicated in the non-financial statement and described with care and in sufficient detail to enable the reader to understand the real impact on the environment,
- comparability – all measurement and estimation methods should be conducted according to uniform principles and in the same cells (units) – any changes in this respect that would occur from period to period must be clearly indicated in the report,
- reliability – the information and processes used in preparing the report should be collected, registered and interpreted in a way that can be

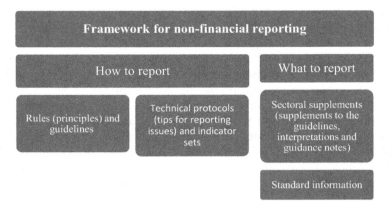

Figure 14.4 Framework for non-financial reporting
Source: Authors' elaboration.

verified and which allows for the determination of their quality and significance,

- transparency – information should be shared and presented in a way that is unambiguous, understandable and available to the reader (stakeholder) of the report,
- documentation (auditability) – all policies, processes, events and results described in the report must be documented in a way that allows for their independent verification,
- timeliness – reports should be prepared regularly and published on dates not later than those provided for by the applicable regulations and should present sufficiently current information to be relevant for readers of reports in their decisions.

Prior to the entry into force of the EU Directive, the literature criticised the lack of a uniform template for non-financial statements (as indicated earlier) and the information disclosed therein that could ensure the comparability of annual reports. It can be stated that the EU Directive imposes a specific layout (structure) for the presentation of non-financial information, which makes the non-financial report more transparent and allows for the comparability of annual reports. Enterprises and institutions which prepare non-financial reports are expected to identify specific thematic aspects and relevant information that should be included in the information they disclose in a reliable, balanced and comprehensive manner, including through cooperation with relevant stakeholders. The information contained in the statement on non-financial information is interrelated. The layout of this information is shown in Figure 14.5.

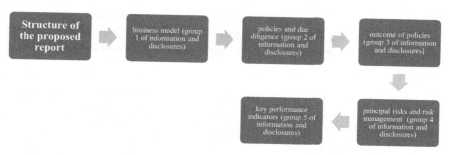

Figure 14.5 Layout of basic information in non-financial reporting
Source: Authors' elaboration.

The first group of basic information is the "business-model" group. The business model of an enterprise or institution describes how it generates and preserves value through its products or services over the longer term. The business model provides context for the management report as a whole and provides an overview of how a company or institution operates and the rationale of its structure by describing how production processes or services are carried out. At this stage, the non-financial report may consider including appropriate disclosures relating to:

- their business environment,
- their organisation and structure,
- the markets where they operate,
- their objectives and strategies,
- main trends and factors that may affect their future development.

To explain their business model, main trends, etc., institutions and companies may consider using performance and effectiveness indicators. The business model must be explained in a clear, understandable and factual manner. The reason for and the moment of changing the business model also require explanation in the reporting year.

The second group refers to refining and presenting the content of the organisation's policies determining its overall commitment to environmental, social and labour issues. Companies and institutions disclose material information that provides a fair view of their policies. In particular, they should disclose their approaches to key non-financial aspects, the main objectives and how they are planning to deliver on those objectives and implement those plans. These explanations should be completed with reference to the specific circumstances of the decision.

The report should include a description of the policies pursued by companies in relation to environmental, social and labour issues; respect for human rights; and the prevention of corruption and bribery. Expanding these areas, it should be stated that the descriptions should concern, among others:

- environmental issues – detailed information on the entity's current and projected environmental impact and, where relevant, health and safety, the use of renewable or non-renewable energy, greenhouse gas emissions, water consumption and air pollution,
- social and employment matters – information presented in the report may cover measures taken to ensure gender equality, implementation of the International Labour Organization's core conventions, working conditions, social dialogue, respect for employees' right to information and expression, respect for trade union rights, health and safety at work and dialogue with local communities or actions taken to ensure the protection and development of these communities, as well as relationships with consumers and responsible marketing and promotion,
- human rights – the statement should contain information on the policies applied in the field of respect for human rights and anti-discrimination,
- counteracting corruption and bribery – the statement may contain information on the measures used.

The non-financial information disclosed by companies and institutions as the outcome of their policies should help stakeholders understand and monitor the company's performance. Relevant disclosures on outcomes of policies may provide useful information on the company's strengths and vulnerabilities and on the performance and effectiveness of their operations. This aspect is the third group presented in the non-financial report (outcome of policies).

The "principal risks and risk management" group is an important group. Companies and institutions should disclose information on their principal risks and on how they are managed and mitigated. Those risks may relate to their operations, their products or services, their supply chain and business relationships. This should also relate to the disclosure of ESG factors (see Table 14.1). It is necessary to show an appropriate perspective in terms of short-, medium- and long-term principal risks.

The key performance indicators (the last group), which are useful given their specific situation, should be reported. Key performance indicators should be consistent with those actually used by the enterprise or institution as part of its internal management and risk assessment processes. The company should disclose key performance indicators that are necessary to understand its development, results, situation and the impact of its activities.

The indicators should reflect the specifics of the industry, sector and business circumstances. Key performance indicators should enable qualitative and quantitative information to be combined and the relationships between them to be outlined.

The disclosure of relevant information should ensure that a balanced and comprehensive picture of the company's development, results and situation, as well as the impact of its business, is obtained. The organisation is expected to provide a clear, reliable and comprehensive picture that covers all relevant aspects of the issue. Table 14.2 presents the basic issues (aspects) related to non-financial reporting. These elements create a non-exhaustive list of thematic aspects that are expected to be considered by organisations when disclosing non-financial information. Comparing the thematic aspects contained in Table 14.2 with Table 14.1 containing aspects in terms of ESG factors, the links and interpenetration between aspects and risk factors are visible.

As the "economic" factor is included in ESG risk, it will not be present in the case of non-financial statements because economic efficiency, including risk, is presented in the financial part of the integrated report, or is a separate report which is presented together with the financial statement. Aspects regarding "social" and "environment" are clearly convergent; however, as indicated, if there is a need to do so, they should be extended. Ethical and human rights-related matters are extended and separate elements.

The success of modern enterprises depends not only on achieving high levels of economic efficiency but also on the social acceptance of their activities. This means that companies need to disclose information on the environmental and social effects of their activities.

Trends and challenges related to non-financial reporting: discussion and conclusions

The impact of business operations is an important consideration when disclosing non-financial information. The impact on the environment can be positive or negative, especially when it comes to issues of sustainability. The disclosure of relevant information should cover both positive and negative types of information in a clear and balanced manner. Organisations can report on a number of potential issues. They assess which information is relevant on the basis of analysing the degree of significance of information for understanding its development, the results and situation and the impact of its activities. Undoubtedly, the analysis of literature and good practices shows that GRI standards have a directional future in applying to include various thematic aspects in a non-financial report because they enable the assessment of significance broken down into internal and external factors.

Table 14.2 Basic issues (aspects) related to non-financial reporting

Employee and respect for human rights	Social	Environmental
Issues of employee • issues of diversity in terms of age, gender, orientation sexual, religion, disability, etc. • participation issues (employee participation and involvement) • conditions for hardships and work • relations with trade unions and respect for the right of freedom of association • human capital management • career path and recruitment management • training/education systems • health and safety at work **Issues of respect for human rights:** • organisation's expectations regarding respect for human rights for employees and business partners • scope of human rights, in particular the rights of employees, temporary employees, children, indigenous communities, owners of small farms, migrants, families of employees • methods of preventing the violation of human rights, including by business partners methods for ensuring repair and remuneration in situations where there has been a violation of human rights	**Issues of social:** • relations with customers (consumers) • satisfaction monitoring • information on the possible health effects of products and security • responsible marketing, promotional activities and research • relations with communities **Issues of anti-corruption:** • policies, procedures, systems and systems to prevent the occurrence of corruption and bribery • training and education in the area of preventing corruption and bribery • operation of the whistleblower mechanism **The materiality of information** • examining the significance of individual non-financial issues • determining whether a given non-financial issue significantly impacts the organization • determining whether the organization significantly affects the data issue • contextual analysis of non-financial issues from the point of view of the organization itself and various stakeholder groups	**Issues of environmental:** • pollution control and prevention • impact on the environment due to energy consumption • direct and indirect emissions to the atmosphere • protection of biodiversity and water sources • waste management impact on the environment due to transport • impact on the environment resulting from the use of services and products and their liquidation **Issues of supply chain (relations with external environment, including contractors):** • providing at least basic information about the structure and structure of the supply chain and about the importance of non-financial information derived from it • in justified cases, information is presented which will allow you to assess how and to what extent the supply chain affects the company's development, performance and position as well as its influence

Source: Authors' elaboration.

A review of the literature shows that the preparation of sustainable financial reporting can be undertaken by organisations, taking into account the theme of sustainability, as well as by fulfilling the obligation arising from applicable legal norms (for example, this fact is indicated by NGFS, 2018; Pereira da Silva, 2019; Ziolo et al., 2019). According to the literature review, non-financial information in the modern economic environment significantly affects the stability of financial systems. ESG reporting has become the norm, and it has been shown that social responsibility issues force recognition of these factors. ESG risk is significantly correlated with the stability of financial systems, as demonstrated in the literature (In et al., 2017; Benlemlih & Bitar, 2018; Ziolo et al., 2019).

The number of companies which disclose their non-financial data is generally increasing. Table 14.2 presents the number of non-financial reports prepared in the GRI standard and registered in the GRI organizations. Table 14.3 provides is a list of reports or published materials registered using the GRI Standards Report Registration System. Data analysis shows a rapid increase in reports. However, those countries in which the number of non-financial reports is falling are marked in grey. There are not significant decreases in the number of reports, albeit this group includes highly developed countries as well as developing countries. Not all EU countries use GRI standards; however, they are used in most countries belonging to this group.

Current trends in reporting on the sustainable development of enterprises are the result of the evolution of the phenomenon which arose in the 1970s. The presented review of the evolution of non-financial reporting forms indicates that applicable law, as well as good practices, indicate flexibility in the scope of data presentation and disclosure. This diversity undoubtedly encourages organisations to take the trouble to present non-financial statements.

The analysis of the data from Table 14.3 not only indicates a growing trend but companies want to improve their market position in an understandable way, and those that seek stability in the long term compile such reports. It should be noted that the number of non-financial reports prepared by financial institutions is growing together with enterprises.

An analysis of selective non-financial reports from the list of GRI Standards report and the pages of EU countries indicates that GRI standards play a dominant role in addition to the EU directive and EU guidelines. Thus, when it comes to non-financial reporting, its evolution, in addition to applicable law, will be influenced by GRI standards and also in terms of expanding the list of disclosures and indicators of scientific work. The implementation of research on the disclosure and impact of ESG factors to be used in building system stability can be observed.

The authors believe that there is a need to examine the impact of non-financial statements on the stability of both the public and market financial

Table 14.3 The number of non-financial reports prepared according to the GRI standard

No.	Country	Belonging to EU countries	2016	2017	2018	2019
1.	Albania					1
2.	Andorra				1	1
3.	Argentina			4	16	23
4.	Australia			12	27	27
5.	Austria	**EU**		7	28	29
6.	Bahrain				1	1
7.	Bangladesh					1
1.	Belgium			8	13	10
2.	Bolivia				3	2
3.	Brazil		1	7	33	44
4.	Cambodia					1
5.	Canada			8	27	41
6.	Chile			2	11	21
7.	Colombia			4	33	56
8.	Costa Rica				4	8
9.	Croatia				4	3
10.	Cyprus	**EU**			1	1
11.	Czech Republic	**EU**			4	6
12.	Denmark	**EU**		1	5	8
13.	Dominican Republic			1		1
14.	Ecuador			1	3	8
15.	Egypt				2	6
16.	Finland	**EU**		5	19	32
17.	France	**EU**		2	5	12
18.	Germany	**EU**		9	52	65
19.	Ghana				9	0
20.	Greece	**EU**		7	21	24
21.	Guatemala				1	2
22.	Honduras					2
23.	Hong Kong			4	27	22
24.	Hungary	**EU**		2	3	4
25.	Iceland					4
26.	India			4	19	21
27.	Indonesia				3	4
28.	Iran				1	1
29.	Ireland	**EU**			3	4
30.	Israel			4	8	7
31.	Italy	**EU**	1	2	39	62
32.	Japan			1	26	24
33.	Jordan			2	2	2
34.	Kyrgyzstan			1	1	0
35.	Kazakhstan			1	2	1
36.	Korea				2	9
37.	Kuwait				4	1
38.	Latvia	**EU**			1	3

No.	Country	Belonging to EU countries	2016	2017	2018	2019
39.	Lebanon			1	3	3
40.	Lithuania	**EU**		1	1	1
41.	Luxembourg	**EU**		5	5	6
42.	Madagascar				1	0
43.	Mainland China			1	1	4
44.	Malaysia				6	10
45.	Mexico			8	20	24
46.	Morocco			1	2	4
47.	Netherlands	**EU**		6	17	18
48.	New Zealand			2	4	3
49.	Nigeria			1	3	2
50.	Norway			2	9	18
51.	Oman				1	1
52.	Pakistan			1	1	1
53.	Palestinian Territories					1
54.	Panama				2	2
55.	Paraguay				1	3
56.	Peru			8	57	149
57.	Philippines			5	8	15
58.	Poland	**EU**		1	7	9
59.	Portugal	**EU**		3	8	16
60.	Qatar			1	1	1
61.	Romania	**EU**			3	4
62.	Russian Federation			5	15	16
63.	Saudi Arabia			2	3	6
64.	Singapore			12	52	51
65.	Slovenia	**EU**			3	1
66.	South Africa			1	1	4
67.	Spain	**EU**	2	18	64	79
68.	Sri Lanca				6	6
69.	Sweden	**EU**	1	9	62	77
70.	Switzerland			7	35	54
71.	Taiwan			3	17	19
72.	Thailand			2	22	35
73.	Turkey			7	15	22
74.	Uganda					3
75.	Ukraine				1	2
76.	United Arab Emirates			4	15	18
77.	United Kingdom of Great Britain and Northern Ireland			6	21	25
78.	USA			28	99	150
79.	Uruguay			1	6	6
80.	Venezuela				1	1
81.	Vietnam					18

Source: Authors' elaboration on list of GRI standards report.

system. While research in this area is conducted at central banks, there is a lack an objective and comprehensive research view of the analysed dependencies. There is a lack of research and positions on whether small enterprises should also prepare non-financial reports or whether it should be reported on a sectoral basis. It is also important that standards continue to evolve and be extended to external factors such as fraud, money laundering and public ethnocentrism.

Disclosures seem very important, especially when we are dealing with crises and the current situation foreshadows the beginning of an economic crisis. Therefore, in the future it will become important to examine whether there is a relationship between non-financial reporting and the stability of financial systems in the face of a crisis.

Note

1 This part was written based on Directive 2014/95/EU, Communication from the Commission – Guidelines on non-financial reporting, Communication from the Commission – Guidelines on non-financial reporting: Supplement on reporting climate-related information (2019/C 209/01), GRI website (Standards GRI).

References

A4S, 2010, *Accounting for Sustainability Newsletter, Accounting for Sustainability*, 4 February 2010, Available online: http://app1.hkicpa.org.hk/correspond ence/2010-02-04/A4S.pdf.

A4S, 2011, *Connected Reporting, a Practical Guide with Worked Examples, Accounting for Sustainability*, Available online: www.accountingforsustainability. org/wp-content/uploads/2011/10/Connected-Reporting.pdf.

A4S, GRI, 2010, *Press Release. Formation of the International Integrated Reporting Committee (IIRC), Accounting for Sustainability*, 2 August 2010, Available online: http://pwc.blogs.com/files/formation-of-the-interna-tional-integrated-reporting -committee.pdf.

AccountAbility 1000, 1999, *A Foundation Standard for Quality in Social and Ethical Accounting, Auditing and Reporting*, London: The Institute of Social and Ethical Accountability, Available online: www.dea.univr.it/documenti/OccorrenzaIns/ matdid/matdid728652.pdf.

Adams, C.A., and Whelan, G., 2009, Conceptualising Future Change in Corporate Sustainability Reporting, *Accounting Auditing and Accountability Journal*, 22, 118–143.

Alińska, A., Filipiak, B.Z., and Kosztowniak, A., 2018, The Importance of the Public Sector in Sustainable Development in Poland, *Sustainability*, 10, 3278.

Ansari, N., Cajias, M., and Bienert, S., 2015, The Value Contribution of Sustainability Reporting – An Empirical Evidence for Real Estate Companies, *ACRN Oxford Journal of Finance and Risk Perspectives*, 4, 190–205.

Arvidsson, S., 2011, Disclosure of Non-Financial Information in the Annual Report a Managementteam Perspective, *Journal of Intellectual Capital*, 12 (2), 277–300.

Astupan, D., and Schönbohm, A., 2012, Sustainability Reporting Performance in Poland: Empirical Evidence from the WIG 20 and MWIG 40 Companies, *Polish Journal of Management Studies*, 12, 68–80.

Basic Guide Communication on Progress. UN Global Compact, 2019, Available online: www.unglobalcompact.org/library/305.

Bebbington, J., and Larrinaga, C., 2014, Accounting and Sustainable Development: An Exploration, *Accounting, Organizations and Society*, 39, 395–413.

Bebbington, J., and Thomson, I., 2007, Social and Environmental Accounting, Auditing, and Reporting: A Potential Source of Organisational Risk Governance? *Environment and Planning C: Government and Policy*, 25, 38–55.

Bebbington, J., Unerman, J., and O'Dwyer, B., 2014, *Sustainability Accounting and Accountability*, New York: Routledge.

Beck, U., 2006, Living in the World Risk Society. *Economy and Society*, 35, 329–345.

Beck, U., and Holzer, B., 2007, Organizations in World Risk Society, in: Pearson, C.M., Roux-Dufort, C., and Clair, J.A., editors, *International Handbook of Organizational Crisis Management*, New York: Sage Publications, pp. 3–24.

Bedenik, N.O., and Barišić, P., 2019, Nonfinancial Reporting: Theoretical and Empirical Evidence, *Sustainable Management Practices, IntechOpen*, doi:10.5772/intechopen.87159.

Benlemlih, M., and Bitar, M., 2018, Corporate Social Responsibility and Investment Efficiency, *Journal of Business Ethics*, 148 (3), 647–671, https://doi.org/10.1007/s10551-016-3020-2.

Blowfield, M., and Murray, A., 2011, *Corporate Responsibility*, New York: Oxford University Press.

Bonacchi, M., and Rinaldi, L., 2007, Dartboards and Clovers as New Tools in Sustainability Planning and Control, *Business Strategy and the Environment*, 16, 461–473.

Boubakary, D., and Moskolaï, D.D., 2016, The Influence of the Implementation of CSR on Business Strategy: An Empirical Approach Based on Cameroonian Enterprise, *Arab Economic and Business Journal*, 11 (2), 162–171.

Bouten, L., and Hoozée, S., 2013, On the Interplay Between Environmental Reporting and Management Accounting Change, *Management Accounting Research*, 24, 333–348.

Breuer, A., Janetschek, H., and Malerba, D., 2019, Translating Sustainable Development Goal (SDG) Interdependencies into Policy Advice, *Sustainability*, 11, 2092.

Burritt, R.L., and Schaltegger, S., 2010, Sustainability Accounting and Reporting: Fad or Trend? Accounting, *Auditing & Accountability Journal*, 23 (7), 829–846.

Cantele, S., 2014, The Trend of Sustainability Reporting in Italy: Some Evidence from the Last Decade, *International Journal of Sustainable Economy*, 6 (4), 381–405.

Carp, M., Pavaloaia, L., Afrasinei, M.-B., and Georgescu, I.E., 2019, Is Sustainability Reporting a Business Strategy for Firm's Growth? Empirical Study on the Romanian Capital Market, *Sustainability*, 11, 658.

CERES, 2004, *Coalition for Environmentally Responsible Economies*. Available online: http://www.ceres.org/.

Chesnes, T., and Newport D., 2000, *University of Florida Sustainability Indicators, August 2001*, The Greening UF Program.

Cho, C.H., and Patten, D.M., 2007, The Role of Environmental Disclosures as Tools of Legitimacy: A Research Note, *Accounting, Organizations and Society*, 32, 639–647.

Clarkson, P., Fang, X., Li, Y., and Richardson, G., 2013, The Relevance of Environmental Disclosures: Are Such Disclosures Incrementally Informative? *Journal of Accounting and Public Policy*, 32, 410–431.

Clarkson, P., Li, Y., Richardson, G., and Vasvari, F., 2008, Revisiting the Relation Between Environmental Performance and Environmental Disclosure: An Empirical Analysis, *Accounting, Organizations and Society*, 33, 303–327.

Communication from the Commission Guidelines on Non-Financial Reporting (Methodology for Reporting Non-Financial Information) (2017/C 215/01), Available online: https://eur-lex.europa.eu/legal-content/EN/TXT/?uri=CELEX:52017XC0705(01).

Communication from the Commission Guidelines on Non-Financial Reporting: Supplement on Reporting Climate-Related Information (2019/C 209/01), Available online: https://eur-lex.europa.eu/legal-content/EN/TXT/?uri=CELEX:52019XC0620(01).

CoP Policy 2015, Available online: www.unglobalcompact.org/library/1851.

Corporate and Sustainability Reporting Trends in Japan, WBCSD, Geneva, 2019.

Crane, S., 2018, *2030. The Future of Sustainability Reporting*, Ottawa: The Conference Board of Canada, Jarislowsky Fraser.

De Villiers, C., Rinaldi, L., and Unerman, J., 2014, Integrated Reporting: Insights, Gaps and an Agenda for Future Research, *Accounting, Auditing & Accountability Journal*, 27, 1042–1067.

Deloitte Touche Tohmatsu Policy on IFRS Terminology 2002, IAS Plus, Published Quarterly, Issue No. 9.

Dhaliwal, D.S., Li, O.Z., Tsang, A., and Yang, Y.G., 2011, Voluntary Nonfinancial Disclosure and the Cost of Equity Capital: The Initiation of Corporate Social Responsibility Reporting, *The Accounting Review*, 86, 59–100.

Dias-Sardinha, I., and Reijnders, L., 2001, Environmental Performance Evaluation and Sustainability Performance Evaluation of Organizations: An Evolutionary Framework, *Eco-Management and Auditing*, 8, 71–79.

Directive 2014/95/EU of The European Parliament and of The Council of 22 October 2014 amending Directive 2013/34/EU as Regards Disclosure of Non-Financial and Diversity Information by Certain Large Undertakings and Groups, Available online: https://eur-lex.europa.eu/legal-content/EN/TXT/?uri=OJ:L:2014:330:TOC.

Ditz, D., Ranganathan, J., and Banks, D., 1995, *Green Ledgers: Case Studies in Corporate Environmental Accounting*, Washington, DC: Wordls Resources Instiyute.

DJSI, 2004, *Dow Jones Sustainability Indexes*, Available online: http://www.sustainability-index.com/.

Dunn, P., 2014, The Impact of Insider Power on Fraudulent Financial Reporting, *Journal of Management*, 30 (3), 397–412.

Eccles, R., and Saltzman, D., 2011, Achieving Sustainability Through Integrated Reporting, *Stanford Social Innovation Review*, 9 (3) (Summer), 56–61.

Epstein, M.J., and Buhovac, A.R., 2014, *Making Sustainability Work: Best Practices in Managing and Measuring Corporate Social, Environmental, and Economic Impacts*, Oakland, CA: Berrett-Koehler Publishers.

Eremina, A.R., Bardadym, M.V., Hurtova, N.V., Zhadko, K.S., and Datsenko, V.V., 2020, Innovative Marketing tools for Business Development in the Early Stages of the Crisis, *International Journal of Management*, 11 (5), 1115–1135.

Falcone, P.M., 2018, Green Investment Strategies and Bank-Firm Relationship: A Firm-Level Analysis, *Economics Bulletin*, 38, 2225–2239.

Falcone, P.M., Morone, P., and Sica, E., 2018, Greening of the Financial System and Fueling a Sustainability Transition: A Discursive Approach to Assess Landscape Pressures on the Italian Financial System, *Technological Forecasting and Social Change*, 127, 23–37.

Falcone, P.M., and Sica, E., 2019, Assessing the Opportunities and Challenges of Green Finance in Italy: An Analysis of the Biomass production Sector, *Sustainability*, 11, 517, doi:10.3390/su11020517.

Farneti, F., and Guthrie, J., 2009, Sustainability Reporting by Australian Public Sector Organisations: Why They Report, *Accounting Forum*, 33, 89–98, Available online: http://isiarticles.com/bundles/Article/pre/pdf/32. pdf.

Figge, F., Hahn, T., Schaltegger, S., and Wagner, M., 2002, The Sustainability Balanced Scorecard – Linking Sustainability Management to Business Strategy, *Business Strategy and the Environment*, 11, 269–284.

Flower, J., 2015, The International Integrated Reporting Council: A Story of Failure, *Critical Perspectives on Accounting*, 27 (3), 1–17.

Friede, G., Busch, T., and Bassen, A., 2015, ESG and Financial Performance: Aggregated Evidence from More Than 2,000 Empirical Studies, *Journal of Sustainable Finance & Investment*, 5, 210–233, https://doi.org/10.1080/20430795.2015.1118917.

FTSE4GOOD, 2004, Available online: http://www.ftse.com/ftse4good/index.jsp.

G4, Sustainability Reporting Guidelines, Available online: www.globalreporting.org/standards/gri-standards-download-center/.

Global Reporting Initiative (GRI), 2002, *Sustainability Reporting Guidelines*, Boston: Global Reporting Initiative.

Gond, J.P., Grubnic, S., Herzig, C., and Moon, J., 2012, Configuring Management Control Systems: Theorizing the Integration of Strategy and Sustainability, *Management Accounting Research*, 23, 205–223.

Gray, R., 1993, *Accounting for the Environment*, London: Paul Chapman.

Gray, R., 1994, Corporate Reporting for Sustainable Development: Accounting for Sustainability in 2000 AD, *Environmental Values*, 17–45.

Hackston, D., and Milne, M.J., 1996, Some Determinants of Social and Environmental Disclosures in New Zealand Companies, *Accounting, Auditing & Accountability Journal*, 9, 77–108.

Hales, J., Matsumura, E.M., Moser, D.V., and Payne, R., 2016, Becoming Sustainable: A Rational Decision Based on Sound Information and Effective Processes? *Journal of Management Accounting Research*, 28 (2).

Hammer, B., and Stinson, C.H., 1995, Managerial Accounting and Environmental Compliance Costs, *Journal of Cost Management* (Summer), 4–10.

Heal, G., 1998, *Valuing the Future: Economic Theory and Sustainability*, New York: Columbia University Press.

Hedberg, C.J., and Malmborg, F.V., 2003, The Global Reporting Initiative and Corporate Sustainability Reporting in Swedish Companies, *Corporate Social Responsibility and Environmental Management*, 10, 153–164.

Henri, J.-F., and Journeault, M., 2010, Eco-Control: The Influence of Management Control Systems on Environmental and Economic Performance, *Accounting, Organizations and Society*, 35, 63–80.

Herzig, C., Viere, T., Schaltegger, S., and Burritt, R.L., 2012, *Environmental Management Accounting: Case Studies of South-East Asian Companies*, London: Routledge.

Holland, H., and Petersen, U., 1995, *Living Dangerously. The Earth, Its Resources, and the Environment*, Princeton: Princeton University Press.

Hristov, I., Chirico, A., and Appolloni, A., 2019, Sustainability Value Creation, Survival, and Growth of the Company: A Critical Perspective in the Sustainability Balanced Scorecard (SBSC), *Sustainability*, 11, 2119.

Hubbard, G., 2009, Measuring Organizational Performance: Beyond the Triple Bottom Line, *Business Strategy and the Environment*, 18, 177–191.

IFRS, 2019, *Conceptual Framework for Financial Reporting and IFRS Practice Statements*, Available online: www.ifrs.org/issued-standards/list-of-standards/.

IIRC, 2017, *Investors Support Integrated Reporting as a Route to Better Understanding of Performance*, September 2017, Available online: http://integratedreporting.org/wp-content/uploads/2017/11/Creating_Value_Benefits toInvestorsWeb.pdf.

In, S.Y., Park, K.Y., and Monk, A.H.B., 2017, *Is "Being Green" Rewarded in the Market? An Empirical Investigation of Decarbonization and Stock Returns*, Stanford Global Project Center Working Paper, Stanford, Stanford University.

ISO, 2004, ISO 14000 [Website online], Available online: www.iso.org/iso/en/iso9000-14000/iso14000/iso14000index.html. Internet.

ISO 26000, Available online: www.iso.org/iso-26000-social-responsibility.html.

ISO, no date, *The ISO Survey of ISO 9000 and ISO 14000 Certificates – Tenth Cycle*, Available online: http://www.iso.ch/iso/en/iso9000-14000/pdf/survey10th cycle.pdf, Internet.

Jonikas, D., 2013, Conceptual Framework of Value Creation Through CSR in a Separate Member of Value Creation Chain, in: Szymańska, D., and Chodkowska-Miszczuk, J., editors, *Bulletin of Geography: Socio Economic Series*, Vol. 21, Toruń: Nicolaus Copernicus University Press, pp. 69–78, http://dx.doi.org/10.2478/bog-2013-0022.

Joshi, S., and Krishnan, R., 2010, Sustainability Accounting Systems with a Managerial Decision Focus, *Cost Management*, 24 (6), 20–30.

Joshi, S., and Li, Y., 2016, What Is Corporate Sustainability and How Do Firms Practice It? A Management Accounting Research Perspective, *Journal of Management Accounting Research*, 28 (2) (Summer), 1–11.

Journeault, M., 2016, The Influence of the Eco-Control Package on Environmental and Economic Performance: A Natural Resource-Based Approach, *Journal of Management Accounting Research*, 28 (2).

Kaplan, R.S., and Norton, D.P., 1996, Linking the Balanced Scorecard to Strategy, *California Management Review*, 39, 53–79.

Kolk, A.A., 2004, Decade of Sustainability Reporting: Developments and Significance, *International Journal of Environment and Sustainable Development*, 3, 51–64.

Kreuze, J.G., and Newell, G.E., 1994, ABC and Life-Cycle Costing for Environmental Expenditures, *Management Accounting* (2) (February), 38–42.

Lamberto, G., 2005, Sustainability Accounting – A Brief History and Conceptual Framework, *Accounting Forum*, 29, 7–26.

Lavorata, L., 2014, Influence of Retailers' Commitment to Sustainable Development on Store Image, Consumer Loyalty and Consumer Boycotts: Proposal for a Model Using the Theory of Planned Behaviour, *Journal of Retailing and Consumer Services*, 21 (6), 1021–1027.

Leipziger, D., 2017, *The Corporate Responsibility Code Book*, London: Routledge.

Lins, C., Althoff, R., and Meek, A., 2008, *Sustainability Reporting in the Mining Sector: Value Association and Materiality*, Brazil: FBDS Fundacao Brasileira, Available online: www.fbds.org.br/fbds/IMG/pdf/doc- 593.pdf.

List of GRI Standards Report, Available online: www.globalreporting.org/reportregistration/verifiedreports.

Loh, L., Thomas, T., and Wang, Y., 2017, Sustainability Reporting and Firm Value: Evidence from Singapore-Listed Companies, *Sustainability*, 9, 2112.

Lourenço, C.I., Callen, J.L., Branco, M.C., and Curto, J.D., 2014, The Value Relevance of Reputation for Sustainability Leadership, *Journal of Business Ethics*, 119, 17–28.

Manes-Rossi, F., Tiron-Tudor, A., Nicolo, G., Zanellato, G., 2018, Ensuring More Sustainable Reporting in Europe Using Non-Financial Disclosure – De Facto and De Jure Evidence, *Sustainability*, 10, 1162.

Mathews, M.R., 1993, *Socially Responsible Accounting*, London: Chapman & Hall.

Mathews, M.R., 1995, Social and Environmental Accounting: A Practical Demonstration of Ethical Concern? *Journal of Business Ethics*, 14 (8), 663–671.

Mazzucato, M., and Semieniuk, G., 2018, Financing Renewable Energy: Who Is Financing What and Why It Matters, *Technological Forecasting and Social Change*, 127, 8–22.

mBank, 2020, *mBank Sustainability Standards*, Available online: www.mbank.pl/pdf/CSR/mbank-sustainability-standards.pdf.

Mensah, J., 2019, Sustainable Development: Meaning, History, Principles, Pillars and Implications for Human Action: Literature Review, *Cogent Social Sciences*, 5, 1653531.

Meyer, J.W., and Rowan, B., 1997, Institutionalized Organizations: Formal Structure as Myth and Ceremony, *The American Journal of Sociology*, 83 (2) (September), 340–363.

Milne, M.J., and Gray, R., 2007, Future Prospects for Corporate Sustainability Reporting, *Sustainability Accounting and Accountability*, 1, 184–207.

Möller, A., and Schaltegger, S., 2005, The Sustainability Balanced Scorecard as a Framework for Eco-Efficiency Analysis, *Journal of Industrial Ecology*, 9 (4), 73–83.

Mulkhan, U., 2013, Corporate Sustainability Reporting: A Content Analysis of CSR Reporting in Indonesia, *Jurnal Perspektif Bisnis*, 1, 73–89.

NGFS, 2018, *First Progress Report*, Paris: Banque de France.

Niño-Muñoz, D., Galán-Barrera, J., and Álamo, P., 2019, Implementation of a Holistic Corporate Social Responsibility Method with a Regional Scope, *Innovar*, 29 (71), 11–30.

North, D.C., 1990, *Institutions, Institutional Change, and Economic Performance*, Cambridge: Cambridge University Press.

O´Dwyer, B., 2002, Managerial Perceptions of Corporate Social Disclosure: An Irish Story, *Accounting, Auditing & Accountability Journal,* 15, 406–436.

OECD Guidelines for Multinational Enterprises, OECD Publishing, 2011, http://dx.doi.org/10.1787/9789264115415-en.

Oh, H., Bae, J., and Kim, S.J., 2017, Can Sinful Firms Benefit from Advertising Their CSR Efforts? Adverse Effect of Advertising Sinful Firms' CSR Engagements on Firm Performance, *Journal of Business Ethics,* 143, 643–663.

Palea, V., 2018, Financial Reporting for Sustainable Development: Critical Insights into IFRS Implementation in the European Union, *Journal,* 42 (3), 248–260, https://doi.org/10.1016/j.accfor.2018.08.001.

Patten, D.M., and Zhao, N., 2014, Standalone CSR Reporting by U.S. Retail Companies, *Account. Forum,* 38, 132–144.

Pearce, D., and Turner, K., 1990, *Economics of Natural Resources and the Environment,* Baltimore: Johns Hopkins Press.

Pereira da Silva, L.A., 2019, *Research on Climate-Related Risks and Financial Stability: An 'Epistemological Break'?* Presented at the Conference of the Central Banks and Supervisors for the Network for Greening the Financial System (NGFS), Paris.

Prado-Lorenzo, J.M., Rodríguez-Domínguez, L., Gallego-Álvarez, I., and García-Sánchez, I.M., 2009, Factors Influencing the Disclosure of Greenhouse Gas Emissions in Companies World-Wide, *Management Decision,* 47, 1133–1157.

Rafiq, M., Zhang, X., Yuan, J., Naz, S., and Maqbool, S., 2020, Impact of a Balanced Scorecard as a Strategic Management System Tool to Improve Sustainable Development: Measuring the Mediation of Organizational Performance through PLS-Smart, *Sustainability,* 12, 1365.

Roth, H., 2008, Using Cost Management for Sustainability Efforts, *Journal of Corporate Accounting & Finance,* 19, 11–18.

Schaltegger, S., and Sturm, A., 1995, *Eco-Efficiency Through Eco-Control,* Zurich, Switzerland: VDF.

Schaltegger, S., and Wagner, M., 2006, Integrative Management of Sustainability Performance, Measurement and Reporting, *International Journal of Accounting, Auditing and Performance Evaluation,* 3, 1–19.

Scott, R.W., 2001, *Institutions and Organizations,* Thousand Oakes, CA: Sage Publications.

Scott, R.W., 2010, Reflections: The Past and Future of Research on Institutions and Institutional Change, *Journal of Change Management,* 10 (1), 5–21.

Seegebarth, B., Peyer, M., Balderjahn, I., and Wiedmann, K.P., 2016, The Sustainability Roots of Anti-consumption Lifestyles and Initial Insights Regarding Their Effects on Consumers' Well-Being, *Journal of Consumer Affairs,* 50 (1), 68–99.

Soll, J., 2015, *The Reckoning. Financial Accountability and the Rise and Fall of Nations,* New York: Basic Books.

Stacchezzini, R., Melloni, G., and Lai, A., 2016, Sustainability Management and Reporting: The Role of Integrated Reporting for Communicating Corporate Sustainability Management, *Journal of Cleaner Production,* 36A (10), 102–110.

Steinmeier, M., 2016, Fraud in Sustainability Departments? An Exploratory Study, *Journal of Business Ethics,* 138, 477–492.

Sutopo, B., Kot, S., Adiati, A.K., and Ardila, L.N., 2018, Sustainability Reporting and Value Relevance of Financial Statements, *Sustainability,* 10, 678, doi:10.3390/su10030678.

Thomsen, C., 2013, Sustainability (World Commission on Environment and Development Definition), in: Idowu, S.O., Capaldi, N., Zu, L., and Das Gupta, A., editors, *Encyclopaedia of Corporate Social Responsibility*, Heidelberg: Springer, pp. 2358–2363.

Thoradeniya, P., Lee, J., Tan, R., and Ferreira, A., 2015, Sustainability Reporting and the Theory of Planned Behaviour, *Accounting, Auditing & Accountability Journal*, 28, 1099–1137.

Tibor, T., and Feldman I., 1996, *Implementing ISO 14000*, New York: McGraw-Hill.

Truant, E., Corazza, L., and Scagnelli, S.D., 2017, Sustainability and Risk Disclosure: An Exploratory Study on Sustainability Reports, *Sustainability*, 9, 636.

Tvaronavičienė, M., Shishkin, A., Lukáč, P., Illiashenko, N., and Zapototskyi, S., 2017, Sustainable Economic Growth and Development of Educational Systems, *Journal of International Business Studies*, 10, 285–292.

Unerman, J., 2000, Methodological Issues – Reflections on Quantification in Corporate Social Reporting Content Analysis, *Accounting, Auditing & Accountability Journal*, 13, 667–680.

Vann, J.W., and White, G.B., 2004, Sustainability Reporting in the Accounting Curriculum, *Journal of Business & Economics Research*, 2 (12), 17–30.

Wang, Y., and Zhi, Q., 2016, The Role of Green Finance in Environmental Protection: Two Aspects of Market Mechanism and Policies. *Energy Procedia*, 104, 311–316.

WBCSD, 2004, *WBCSD's SD Reporting Portal*, Available online: www.wbcsd.org/includes/getTarget.asp?type=e&id=www.sdportal.org/templates/Template3/layout.asp?MenuId=360.

Wensen, K., Broer, W., Klein, J., and Knopf, J., 2011, *The State of Play in Sustainability Reporting in the European Un ion 2010. Final Report*, Available online: http://ec.europa.eu/social/main.jsp?langId=en&catId=89&newsId=1013.

Young, C.W., Hwang, K., McDonald, S. et al., 2010, Sustainable Consumption: Green Consumer Behaviour When Purchasing Products, *Sustainable Development*, 18 (1), 18–31.

Ziolo, M., Filipiak, B.Z., Bąk, I., Cheba, K., Tîrca, D.M., and Novo-Corti, I., 2019, Finance, Sustainability and Negative Externalities. *An Overview of the European Context, Sustainability*, 11, 4249.

Index

Printed in the United States
By Bookmasters